Fieldwork in the Global South

T0298715

Choosing to do fieldwork overseas, particularly in the Global South, is a challenge in itself. The researcher faces logistical complications, health and safety issues, cultural differences, language barriers, and much more. But permeating the entire fieldwork experience are a range of intermediating ethical issues. While many researchers seek to follow institutional and disciplinary guidelines on ethical research practice, the reality is that each situation is unique and the individual researcher must negotiate their own path through a variety of ethical challenges and dilemmas. This book was created to share such experiences, to serve not as a manual for ethical practice but rather as a place for reflection and mutual learning.

Since ethical issues face the researcher at every turn and cannot be compartmentalized into one part of the research process, this book puts them at the very center of the discussion and uses them as the lens with which to view different stages of fieldwork. The book covers four thematic areas: ethical challenges in the field; ethical dimensions of researcher identity; ethical issues relating to research methods; and ethical dilemmas of engagement with a variety of actors. This volume also provides fresh insights by drawing on the experiences of research students rather than those of established academics. The contributors describe research conducted for their master's degrees and doctorates, offering honest and self-critical reflections on how they negotiated ethical challenges and dilemmas.

The chapters cover fieldwork carried out in countries across Africa, Asia, and Latin America on a broad sweep of development-related topics. This book should have wide appeal to undergraduates, postgraduates, and early-career researchers working under the broad umbrella of development studies. Although focused on fieldwork in the Global South, the discussions and reflections are relevant to field research in many other countries and contexts.

Jenny Lunn manages a project promoting public engagement in geography at the Royal Geographical Society (with the Institute of British Geographers) in London.

Routledge Studies in Human Geography

This series provides a forum for innovative, vibrant, and critical debate within Human Geography. Titles will reflect the wealth of research which is taking place in this diverse and ever-expanding field. Contributions will be drawn from the main sub-disciplines and from innovative areas of work which have no particular sub-disciplinary allegiances.

Published:

Fieldwork in the Global South

Ethical challenges and dilemmas

Edited by Jenny Lunn

Routledge
Taylor & Francis Group

LONDON AND NEW YORK

First published 2014
by Routledge
2 Park Square, Milton Park, Abingdon, Oxon OX14 4RN

and by Routledge
52 Vanderbilt Avenue, New York, NY 10017

First issued in paperback 2020

Routledge is an imprint of the Taylor & Francis Group, an informa business

British Library Cataloguing in Publication Data
A catalogue record for this book is available from the British Library

Library of Congress Cataloging in Publication Data
Lunn, Jenny. Fieldwork in the global south/Jenny Lunn.

pages cm
Includes bibliographical references and index.
1. Ethnology–Fieldwork–Southern Hemisphere. 2. Anthropological ethics–Southern Hemisphere. 3. Participant observation–Moral and ethical aspects–Southern Hemisphere. 4. Southern Hemisphere–Social life and customs. I. Title.
GN590.L95 2014
305.8009181′4–dc23

2013036608

ISBN 13: 978-0-367-66958-4 (pbk)
ISBN 13: 978-0-415-62841-9 (hbk)

Typeset in Times New Roman
by Wearset Ltd, Boldon, Tyne and Wear

Contents

x *Contents*

Contributors

Arshiya Bose completed a degree in Liberal Arts studying Biology and Creative Writing at Bryn Mawr College, USA. Prior to her postgraduate studies she worked with Kalpavriksh Environment Action Group on policy and advocacy relating to conservation and livelihood issues concerning forest-dwelling communities in India. She then studied for a master's degree and doctorate in the Department of Geography at the University of Cambridge. Her doctoral research explored the dynamics between people's use of forests and biodiversity conservation interventions. Her PhD fieldwork in the Western Ghats biodiversity hotspot in India threw her headfirst into the world of coffee, through both consumption and production. She currently works on conserving biodiversity in coffee landscapes in the same region.

Andrew Brooks began his academic studies at King's College London with a BA in Development Geography. He then worked for two years for Voluntary Services Overseas, first as an Education Researcher with the Department of Education in Papua New Guinea where he investigated the impact of the introduction of a new primary school syllabus, then as a Goal Support Officer contributing to International Program Development in the UK office. He then studied for an MSc in Practising Sustainable Development followed by a doctorate in the Geography Department at Royal Holloway, University of London. His PhD examined the transnational second-hand clothing trade with fieldwork carried out in Mozambique. He is now a Lecturer in Development Geography at King's College London. His research focuses on the political economy of southern Africa and investigates connections between places of consumption and spaces of production, and commodity chains which link the Global North and South.

Margi Bryant graduated from New Hall, University of Cambridge with a degree in Archaeology and Anthropology. She worked for a number of years in journalism, mostly in the Global South, and later moved on to communications work for international development and environment organizations. More recently she was involved in consultancy work on the 'people-facing' side of environmental management. She then joined the Department of Geography at the University of Sheffield for an MA in Social and Cultural

Geography followed by a doctorate. Her research focused on environmental conservation and international volunteering, with fieldwork carried out in Kenya.

Deepta Chopra did her undergraduate training in Economics in India. She then studied for a master's degree in Social Work (India's equivalent to Development Studies). After working for a few years in a small rural community in Bastar (Chhatisgarh), India she moved to the UK to study for a master's degree in Development Studies at the Institute of Development Studies, University of Sussex. Following a short period of work at the Working Lives Research Institute at London Metropolitan University she moved to the Department of Geography at the University of Cambridge to pursue her doctorate. Her research examined the political economy of the formulation and early implementation of the Mahatma Gandhi National Rural Employment Guarantee Act in India. Most of her fieldwork was carried out in Delhi and Rajasthan. Since completing her PhD, she has been working as a Research Fellow at the Institute of Development Studies, Sussex.

Rinita Dam's undergraduate degree in Biomedical Sciences at Sheffield Hallam University included a placement year at Pfizer pharmaceuticals. She subsequently studied for a master's degree in Reproductive and Sexual Health Research at the London School of Hygiene and Tropical Medicine. Recognizing that she had a passion in 'health rights' but did not have any experience in the social sciences, she travelled to Kolkata, India, in order to do voluntary work with a rights-based sex workers' organization. Following this she studied for a doctorate in the International Development Department at the University of Birmingham. Her PhD research examined the impact of HIV/AIDS on the lives of the poor, specifically exploring the coping strategies of individuals and households and their access to treatment and support services. Her field research was carried out in Kolkata. After completing her PhD she moved back to Asia to carry out research and consultancy work relating to reproductive and sexual health.

Caroline Day's undergraduate degree was in Geography and Development Studies at the University of Sussex. She then studied for an MA in Children, Youth and International Development at Brunel University in London including a dissertation on the representation of street children in Kenya. She has held research posts at the UK charities Barnardo's and Centrepoint working with vulnerable children and young people from a variety of social, cultural, and economic backgrounds. She has also volunteered and conducted research in sub-Saharan African countries including Uganda, Kenya, Tanzania, and Zambia, and has been a youth worker in the UK. She then started a doctorate in the Department of Geography and Environmental Science at the University of Reading. Her research focused on the life transitions of young people aged 14 to 30, particularly those with caring responsibilities for chronically ill or disabled parents or relatives, and her fieldwork was carried out in Zambia.

After completing her PhD, Caroline plans to continue researching within the field of the geographies of children, youth, and families, working with young people in the UK as well as continuing her work in Zambia.

Nora Fagerholm studied for a BSc and MSc in Geography and Environmental Sciences in the Department of Geography at the University of Turku, Finland. Having developed an interest in participatory methods and geographic information systems (GIS) she remained in the same department for a doctorate. Her research examined the potential of local community participation in spatial planning and landscape management. She carried out fieldwork in Zanzibar, Tanzania using participatory GIS techniques to create place-based local knowledge on landscapes. She also studied the changes in tropical forest landscapes and their spatial patterns over time. Her postdoctoral research continues to explore participatory GIS methodologies in both Zanzibar and Finland.

Danielle Gent studied for a BSc in Geography and an MSc in International Financial and Political Relations, both at Loughborough University. She also gained a Diploma in International Studies during an Erasmus exchange at the University of Valencia, Spain. She returned to the Department of Geography at Loughborough University for her doctorate. Her research investigated the way in which global developments in renewable energy influenced the direction of solar energy projects in Nicaragua. She is currently a Research Associate in the same department and her research continues to focus on the governance of sustainable energy in the Global South. She also works in a voluntary capacity on various projects in Honduras and Nicaragua.

Girija Godbole studied for a BSc in Botany and an MSc in Anthropology, both at the University of Pune, India. She then worked for about ten years with grassroots organizations and research institutions in the field of rural development and natural resource management. She moved to the Department of Geography at the University of Cambridge to study for an MPhil in Environment, Society and Development, followed by a doctorate. Her research examined the relationship between women and the land in a rapidly changing rural society in western India. After finishing her PhD, she plans to combine research and teaching, ideally conducting field-based research in the Global South with institutional affiliation in the West.

Stephen Jones completed an undergraduate degree in Civil Engineering at Cambridge University. He then worked as a water and sanitation engineer in Kyrgyzstan and was Chief Executive of Engineers Without Borders UK. He moved to the Department of Geography at Royal Holloway, University of London where he studied for a master's degree in Sustainable Development followed by a doctorate in collaboration with the charity WaterAid. His research investigated the challenge of developing flexible financing arrangements which enable sustainable access to water and sanitation for the poor. His fieldwork was carried out in Mali in conjunction with WaterAid.

He currently works in the Democratic Republic of Congo as the water, sanitation and hygiene monitoring and evaluation coordinator for the DRC WASH Consortium.

Riamsara Kuyakanon Knapp grew up and went to school in Thailand; due to her cross-cultural heritage she has been travelling between 'East' and 'West' all her life. She read for an undergraduate degree in Classics and English Literature at the University of Edinburgh and then spent a decade living in various countries and working in different fields including wildlife conservation. She studied at the University of Cambridge for an MPhil in Environment, Society and Development followed by a doctorate. Her research examined the relationship between environmental conservation, development, and Buddhist culture in Bhutan. After her PhD she intends to continue working in the nexus of conservation, culture, and development.

Eszter Krasznai Kovàcs started out in Ecology research at the University of Sydney, Australia, initially studying small marsupial behavior in the forests north of the city. After trapping for months at dawn, she realized that the human side of conservation might be more her true calling. She completed a master's degree in Environmental Law at the University of Sydney and then worked in public interest environmental law. She next moved to the Department of Geography at the University of Cambridge for an MPhil and doctorate. Her research focused on the evolution of the conservation agenda within agricultural areas in the European Union, using Hungary as a case study for the implementation of policies. She plans to continue combining her interest in environmental law with academia.

Virginie Le Masson has an engineering background with a Maîtrise in Development and Land Use Planning from the University of Grenoble 1, France. She moved to Northumbria University in the UK for an MSc in Disaster Management and Sustainable Development followed by a doctorate based in the Centre for Human Geography at Brunel University in London. Her research compared the relevance of climate-related interventions and disaster risk reduction strategies to local experiences of environmental changes. Her fieldwork was carried out in Ladakh, Northern India. Following her PhD she plans to work for NGOs involved in disaster risk reduction projects.

Hayley Leck studied for a Bachelor of Social Science degree in Environmental Studies followed by an Honours degree in Geography and Environmental Management, both at the University of KwaZulu Natal, South Africa. She then worked in various environmental positions in both South Africa and the UK including as an environmental advisor on a gas pipeline project in West Yorkshire, England. She later moved to the Department of Geography at Royal Holloway, University of London, for a doctorate. Her research explored the barriers and opportunities for adaptation to climate change at municipal and household levels in the greater eThekwini (Durban) Metropolitan area, South Africa. After her PhD, she joined the Grantham Research

Institute on Climate Change and the Centre for Climate Change Economics and Policy, London School of Economics and Political Science, as a Postdoctoral Researcher. She continues to be involved in research on climate change adaptation and the political economy of climate resilient development.

Jenny Lunn studied for a BA in Geography at Durham University. Following a year in India working for a variety of NGOs she then pursued an MA in the Geography of Third World Development at Royal Holloway, University of London including fieldwork in Kolkata, India. After several years of working for both charities and consultancies she returned to the Department of Geography at Royal Holloway for a doctorate. Her research examined the role of religious organizations in development practice including the beliefs and scriptures that motivate people of different faiths to engage in activities related to poverty alleviation with fieldwork in Kolkata. After her PhD, she returned to the Royal Geographical Society (with the Institute of British Geographers) in London, where she had worked for over three years prior to her doctoral research, and is involved in a project which promotes public engagement in geography.

Alessandro Moscuzza studied for a BA in Geography and European Studies at London Guildhall University followed by an MA in the Geography of Third World Development at Royal Holloway, University of London including field research in Barbados. He then joined the UK Government's Department for International Development (DFID). He subsequently studied for an MSc in Environmental Management and Protected Areas Management at Birkbeck, University of London with fieldwork carried out in Italy. He is currently an Environment and Climate Change Advisor in DFID's Research and Evidence Division and has had overseas postings in Sierra Leone, in India and with the European Commission in Brussels.

Kamakshi Perera-Mubarak's undergraduate studies were in Geography and Environmental Studies at the University of Melbourne, Australia. Following this she was involved in voluntary work in community development at various NGOs in Sri Lanka, Australia, and the UK before working at MAS Holdings (Pvt) Ltd Sri Lanka in the field of corporate social responsibility. She then moved to the School of Geography and the Environment at the University of Oxford for a doctorate. Her research explored post-tsunami livelihood recovery in Sri Lanka including practices of corruption, political interference, and marginalization of women. She currently works for the Ministry of Defence and Urban Development in Sri Lanka as a Social Specialist in the Metro Colombo Urban Development Project.

Chloe Skinner read for a BSc in Geography and then an MA in International Development, both in the Department of Geography at the University of Sheffield. She remained in the same department as part of the White Rose 'Transformative Justice' research network for her doctorate. Her research explored gender and violence in Palestine and Israel with fieldwork carried out in East Jerusalem and the West Bank. After her PhD she intends to remain in the sphere of feminist and peace research and activism.

Thomas Aneurin Smith completed a BA and a PGCE at Southampton University and then spent some time teaching Geography at secondary school level. He later went on to study for an MRes in Human Geography and then a doctorate at Glasgow University. After a short spell of teaching in the School of Social Justice and Inclusion at the University of Wales Trinity Saint David in Carmarthen, he moved to the Department of Geography at Sheffield University. His research has focused on environmental education, local knowledges of environmental issues, participation in local development projects, the agency of children and young people in the Global South, and the role of non-governmental organizations in education-themed development projects. The majority of his field research has been carried out in Tanzania.

Samantha Staddon studied for an undergraduate degree in Ecology at Lancaster University and then a master's degree in Biodiversity and Conservation at Leeds University. Following this she worked for seven years in applied conservation and academic research in Peru, Tanzania, the Cook Islands, New Zealand, Scandinavia, and the UK. She then decided to change her focus to the people inherent in any attempt to manage nature. She studied for a master's degree in Natural Resource Management and then a doctorate in the School of Geosciences at the University of Edinburgh. Her research examined the involvement of local communities in the monitoring and management of natural resources, with fieldwork carried out in the community forests of Nepal. She now works as Teaching Fellow in Environment and Development at the University of Edinburgh.

Luke Taylor completed a BA in Urban Studies and Planning at the University of Sheffield with a year in Illinois, USA. Following this he studied for an MA in International Development and Planning, also at the University of Sheffield. His research explored the perceptions of tenure security within squatter communities in Jamaica. He then completed a placement at the Global Land Tools Network (GLTN) within the UN-HABITAT Headquarters in Nairobi, Kenya and worked with several NGOs in sub-Saharan Africa on access to land and land rights of the poor. Having completed a short spell of teaching in London, he returned to planning and development and is currently working as a Town Planning Policy Advisor in the UK Government's Department for Communities and Local Government (DCLG).

Julia Tomei studied for a BSc in Biology (International) at the University of Leeds and then an MSc in Environmental Technology (with Ecological Management) at Imperial College London, with a thesis on agricultural biodiversity in the Peruvian Andes. Following this she spent four years working as a Research Associate at the Policy Studies Institute and then at University College London's Energy Institute on a wide range of projects covering sustainable consumption, public engagement with renewable energy, and the sustainability of bioenergy and hydrogen systems. She then enrolled for a doctorate at the UCL Energy Institute. Her research focused on the global governance of biofuels and how this shaped the political economy of biofuels in Guatemala. Since finishing her PhD she has remained at the UCL Energy Institute as a Research Associate and her research interests concern the social and environmental dimensions of energy, particularly the impacts of bioenergy production, processing, and use in developing-world producer countries.

Yue Wang obtained a BA in Economics at Donghua University, China and then an MSc in Human Geography at the Chinese Academy of Sciences. She spent two years working as a business consultant in a foreign-owned company in Shanghai before moving to the School of Environment and Development at the University of Manchester for a doctorate in Geography. Her research examined how the procurement activities of leading retail transnational corporations were changing local supply networks. She carried out fieldwork in her home city of Shanghai. Following her PhD she returned to China as a Research Fellow at the Nanjing Institute of Geography and Limnology, Chinese Academy of Sciences.

Katie Willis studied at Oxford University for a BA in Geography followed by an MPhil in Latin American Studies and a DPhil in Geography. Her postgraduate fieldwork was based in Oaxaca City, Mexico and focused on gender, class, employment, and social networks. She was a Lecturer in Human Geography at the University of Liverpool from 1994 and then moved to Royal Holloway, University of London in 2003. She is currently Professor of Human Geography and Director of the Politics, Development and Sustainability Group within the Geography Department at Royal Holloway. Her research focuses largely on gendered and household-level changes within the context of 'development,' with field research mainly conducted in Latin America (especially Mexico) and East Asia (China). She is the author of two widely-used development textbooks: *Theories and Practices of Development* (2nd edn 2011, Routledge) and *Geographies of Developing Areas: the Global South in a changing world* (2nd edn with Glyn Williams and Paula Meth, 2014, Routledge).

Preface

Choosing to do fieldwork overseas, particularly in the Global South, is a challenge in itself. Travel and logistical complications, health and safety issues, cultural differences, and language barriers are just some of the obstacles that face the researcher, but permeating the entire fieldwork experience are a range of intermediating ethical issues. While researchers seek to follow guidelines on ethical research practice, the reality is that each situation is unique and the individual researcher must negotiate their own path. *Fieldwork in the Global South: Ethical Challenges and Dilemmas* was created to share such experiences, to serve not as a manual for ethical practice but rather as a place for reflection and mutual learning.

The idea for this book evolved out of two events organized by postgraduate committee members of the Developing Areas Research Group (DARG). The first was a workshop on 'Sharing fieldwork experiences' held in London in April 2010, which provided a forum for postgraduates to discuss and reflect on various aspects of overseas fieldwork. Ethical issues emerged as a recurrent theme throughout the workshop; in fact it seemed to be the 'hot topic' to which every discussion returned. In response we organized a session at the Annual Conference of the Royal Geographical Society with the Institute of British Geographers, held in London in September 2010, specifically focusing on 'Ethical challenges and dilemmas of fieldwork in the Global South.' The room was packed tight – not only with postgraduate students, but also with well-established academics – and there was a long and lively debate after the papers had been presented. This highlighted the pertinence and endurance of these debates about the ethical dimensions of field research, yet at the same time the lack of set procedures, neat solutions, or fixed answers. Thus we decided to think about drawing together some of the papers from the conference session for publication. Proposals rolled in from other fellow postgraduates all eager to share their experiences on this complex and unresolved topic.

We wanted this book to be distinctive in two ways. The first was to put ethical challenges and dilemmas at the very center of discussions. Many existing guides on development fieldwork present ethical issues as a single stand-alone chapter, as if ethics could be dealt with separately from the rest of the research cycle. The reality, however, is that ethical issues face the researcher at every turn and cannot be compartmentalized as one discrete part of the research process. Thus we use ethical issues as a lens through which to view different stages of field-work. The book is divided into four thematic sections: ethical challenges in the field; ethical dimensions of researcher identity; ethical issues relating to research methods; and ethical dilemmas of engagement with a variety of actors. Each chapter examines how a particular ethical issue has affected a researcher's fieldwork.

The second distinctive feature of this book is that its contributors were all postgraduate students at the time of writing. Many of the existing guides on development fieldwork have been written by established academics with many years of experience. In contrast we offer fresh insights. All chapters describe fieldwork conducted for master's degrees and doctorates and were written either during the period of study or very shortly after. Each contributor recounts their personal experiences of negotiating particular ethical challenges and dilemmas during their fieldwork, using stories, anecdotes, and self-critical reflections. Each chapter is self-standing so that the book can be dipped into or read systematically.

I am enormously grateful to all of the contributors who have taken time to write a chapter for this book amid their other commitments – some while still overseas on fieldwork, others while in the middle of writing up their doctoral thesis; some having just started a new job, others while starting a new life as a parent. I particularly value their willingness to be honest about their experi-ences – including difficulties, problems, uncertainties, and failures – and to explore their self-critique in a public and published forum in order to help future generations of researchers to negotiate similar ethical challenges and dilemmas.

I am particularly grateful to Katie Willis from Royal Holloway, University of London, who has kindly written an Afterword. She and Elsbeth Robson jointly edited *Postgraduate Fieldwork in Developing Areas: A Rough Guide* (1994, 2nd edn 1997) when they were young academics. The monograph, pub-lished by the Developing Areas Research Group, has been an invaluable refer-ence for students ever since. Much of their volume remains relevant, although some things have changed in the 20 years since Katie and her colleagues drew on their own doctoral and early-career fieldwork experiences, including access to the internet and mobile phones, an increasing institutional concern with health and safety and risk assessments, and a greater number of students from the Global South at Western universities returning to do research 'at home.' It is to be hoped that this book by a new generation of DARG postgraduates, focusing specifically on the ethical dimensions of fieldwork, is both com-plementary and a worthy successor.

I would also like to thank Glyn Williams from the University of Sheffield, who was serving as the chairman of DARG when this book was being prepared, for his advice and encouragement at various stages of the project. Appreciation also goes to my parents – who thought that my doctoral thesis was the last piece of academic proof-reading that they would be obliged to do – for reading all the chapters and providing helpful feedback. Grateful thanks, too, to Lucy Metzger for her meticulous copy-editing of the text in its final stages.

Finally a word of thanks to fellow students with whom I shared part of the postgraduate journey, meeting over coffee (or something stronger!) to discuss, console, share, moan, and encourage. These interactions were invaluable, and I hope that this book acts as a lasting legacy of such conversations and can benefit many more people.

Jenny Lunn
August 2013

1 Rethinking ethics in field research

Integral, individual, and shared

Jenny Lunn

The aim of *Fieldwork in the Global South: Ethical Challenges and Dilemmas* is to provide an accessible resource on some of the complexities and tensions surrounding ethical behavior and practice faced during field research, specifically under the umbrella of development studies. This compilation of experiences and reflections from people who have conducted fieldwork predominantly in Asia, Africa, and Latin America makes the case for rethinking the way in which we view ethics in field research in three ways: by framing ethics as an integral part of the research process; by exploring ethics as something that is experienced individually; and by sharing personal stories as a way of learning from one another.

Many existing textbooks and guides on doing fieldwork in the Global South present ethical issues as a single self-standing section alongside chapters about preparation and logistics, culture shock, health and safety, research methods, language, data collection, and dissemination. The reality, however, is that ethical issues permeate every aspect of the research process: before, during and after fieldwork; when 'at work' as well as during time off; in relationships not only directly with research participants but also with host communities, gatekeeper organizations, institutions, funders and governments, as well as with partners, family, and friends. Each of the chapters in this volume describes how ethical considerations relate to a particular aspect of field research, from the choice of fieldwork location to research methodology and from dealing with vulnerable people to researcher identity. By putting ethics at the heart of every discussion, this volume shifts away from considering ethical issues as a compartmentalized and discrete part of the research process to accepting them as inherent and integral.

Guidelines for ethical research are produced by academic institutions, funding bodies, and disciplinary organizations. While these are well-intentioned in aiming to encourage good practice, the guidelines tend to be somewhat idealized and detached from the reality in the field. Ethical issues are rarely simple or straightforward enough to be dealt with by adhering to a set of procedures; rather they are complicated and sometimes conflictual. The researcher must face problems, negotiate positions, make careful judgments, and take difficult decisions, each of which is unique to its own particular context. Each of the chapters in this volume is a personal account of the ethical challenges and dilemmas faced during fieldwork and of how they were negotiated and not necessarily resolved. By presenting ethics as an individualized experience, this volume challenges the appropriateness of the institutional and procedural approach to ethical research practice.

Although the experience of ethical issues in field research is subjective, there are advantages in sharing reflections with one another. This is not another text-book or manual but rather a collection of personal experiences. It contains stories and anecdotes from fieldwork, honest reflections and self-critique. Some of these are amusing, illustrating how fieldwork can be dynamic and fun; others are upsetting, highlighting the personal impact and frustrations of time spent in the field. Furthermore, rather than representing the voice of academics with many years of experience, all of the chapters have been written by relative novices on their first or second periods of extended individual fieldwork carried out for master's degrees and doctorates. Each contributor grapples with the question, 'did I do the right thing?' It is hoped that the honest tone will provide reassurance and advice, as well as stimulating further debate.

Ethics in field research

Ethics is a branch of philosophy that has been discussed since antiquity. The fact that it remains hotly debated today is indicative of its complexity and of its con-tested and evolving nature. At the core of ethics lie debates about human behav-ior, specifically how people *should* act, or what is 'morality.' Put simply, when faced with a particular situation, the way in which a person reacts, responds, or performs involves ethics. Behavior which deviates from the accepted norm may be deemed 'unethical' or 'immoral.' The complexity derives from the lack of universal absolutes in terms of what is ethical or unethical. While ethics is a matter of much intellectual debate, it is also very much an applied field. Ethics is relevant across daily human interactions and social life, from politics to medi-cine, from law to the military, and from business to research. In research it is particularly pertinent to that which involves other human beings, known as 'human subject research,' an area that spans the medical and social sciences. This volume is not concerned with medical research or health sciences but rather focuses on ethics in relation to social science field research, specifically that which is conducted in the Global South under the broad umbrella of develop-ment studies.

The fundamental concern of development studies is to analyze and better understand the economic, political, social, cultural, and environmental contexts of countries of the Global South and to use that knowledge to address global inequalities and improve the life chances of the poor and marginalized. This volume comprises research from across Africa, Asia, and Latin America on a variety of topics including housing, employment, health and wellbeing. Although there is a strong focus on people there is also a concern with the infrastructure that supports human society, including natural resources, energy, transport, political systems, and so on. Conducting field research in the Global South does entail some practical and ethical challenges which, while not necessarily unique to the Global South, may be heightened in that context. These include working in countries which are politically unstable and dangerous for travel; dealing with governments that are not transparent and where corruption is rife; the paucity of base information for research, such as detailed census data or accurate maps; the challenges of working across cultures in a different language and at a different pace of life; and accessing people who are socially marginalized or dispossessed and understanding the reality of their lives. In such potentially difficult fieldwork contexts, ethics permeates every aspect of the research project. Ethical considerations influence the planning and preparation; they are vitally important in the field at all times, both during data collection and during time off; and they are significant in terms of writing and presenting the findings in written or oral formats. Whether the researcher is an 'outsider' in their chosen fieldwork location or an 'insider' doing research 'at home,' they face ethical challenges at every turn.

As already outlined, ethics concerns appropriate behavior: specifically behavior toward other human beings, whether as individuals, groups, or organizations. I suggest that ethics in fieldwork revolves around three main factors: (1) the *identity* of the researcher and how this affects their positionality, their relationship with the research community, and their data collection; (2) the *behavior* of the researcher in dealing with people, organizations, and data; and (3) the *legacy* of the researcher, including 'giving back' and the representation of data and voices. Since ethical issues relate to every relationship, every encounter, and every action, they are multi-directional, multi-layered, and multi-temporal. As Samantha Staddon says in this volume, she faced 'an ongoing barrage of ethical choices.' Thus ethics simply cannot be considered as a self-standing or discrete part of the research process but rather should be thought about in a more integrated way.

Limitations of institutional ethical guidelines

Before conducting research – whether in a laboratory, archive, classroom, or foreign country – most researchers must undergo an ethical review. This requires them to describe the proposed research and to demonstrate how it will adhere to ethical principles. In most academic institutions, this is reviewed by a committee who decide whether to grant approval or request some alterations. Such institutional concern over ethical research practice is actually a relatively recent phenomenon. It first emerged in the field of medicine after centuries of what is now considered unethical behavior, such as the testing of vaccines, medicines, and biological weapons on slaves, prisoners of war, and children without their consent. The Nuremberg Code of 1948 was the first international document to draw up some ethical guidelines to govern the use of human subjects in research. The first of its ten points states that 'the voluntary consent of the human subject is absolutely essential' ('The Nuremberg Code' 1996). Since then, ethical procedures and safeguards have improved in medical ethics and have spilled over into other research involving human subjects across the sciences and social sciences.

Nowadays almost every academic institution, funding body, and disciplinary organization has a set of ethical guidelines drawn up for their community or users. For example, the 'broad principles' of the Developing Areas Research Group are that:

> Members of DARG should endeavour to incorporate the following broad principles in their work in and on the developing world: honesty, integrity, sensitivity, equality, reciprocity, reflectivity, morality, contextuality, non-discriminatory, fairness, awareness, openness, altruism, justice, trust, respect, commitment.
>
> (DARG 2009)

While it is laudable that institutions, funders, and disciplines encourage their researchers to follow good practice, there are a number of problems with this approach that have been identified by many of the contributors to this book as well as by others.

The first limitation is the broad application of ethical guidance across different areas of academic research. The ethical frameworks set by many academic institutions and funding bodies are designed to be universally applied to all researchers, but this does not always allow enough flexibility to account for contrasting epistemological and methodological approaches. For example, can the subjects in a medical study be considered in the same way as the participants in an ethnographic study? Fixed procedures such as printed consent forms, which might be absolutely vital for a clinical trial, may be wholly inappropriate for illiterate villagers.

A second criticism is the way in which ethical review is proceduralized. It is a form to be filled out in order to comply with institutional requirements; a bureaucratic hoop to jump through in the early part of a research project; a task to be checked off the list along with other preparatory tasks. This procedural approach treats the consideration of ethics as a self-standing, one-time event, but the reality is that ethical challenges crop up throughout the research process and constantly need negotiation and renegotiation, flexibility, and sensitivity to local contexts. However, when a researcher faces changed circumstances in the field they are not required to pause their research and resubmit an ethics review.

A third weakness concerns the question of whom ethical guidelines are in place to protect. They are generally focused on the behavior of the researcher toward research participants and the data collected from them, revolving around the core concepts of informed consent, 'do no harm,' right to privacy, anonymity, confidentiality, and so on. However, little attention is paid to how these principles relate to the researcher. Their health, safety, and wellbeing largely fall under the remit of institutional 'risk assessments,' which cover everything from safe travel to sunburn; but, as several contributors to this volume highlight, the researcher can face potential and actual harm – in some cases physical, in other cases emotional – that is not really covered by ethical guidelines or risk assessment.

A fourth way in which ethical guidelines have been criticized concerns the actual terms used. Many concepts are abstract, such as 'integrity,' 'justice,' and 'transparency,' and do not readily translate into everyday interactions in the field. Many are also traits of personality which are interpreted subjectively rather than being absolute, universal standards. In some ways ethical guidelines set an ideal of the researcher as a perfect moral human being, but Thomas Smith in this volume suggests that they are 'mostly unattainable' for an 'ordinary human being.'

Thus while ethical guidelines and the ethical review process do encourage researchers to think through the ethical implications of their research and to take responsibility for them, the institutional and procedural approach does not always help in negotiating the everyday realities in the field – the gray areas, complex relationships, changing circumstances, subjective feelings, and individuality of the experience. As Sultana (2007: 379) writes, ethical choices 'are not captured in the "good" ethical guidelines of institutional paperwork, but have to be negotiated and grappled with on a daily basis in the field.' The institutional approach also frames ethics as issues to be identified in advance and mechanisms put in place to safeguard against harm. The reality, however, is not as simple as a problem-and-solution approach. As Chloe Skinner reflects in this volume:

> Many of my ethical dilemmas remain unresolved. I am painfully aware of the 'gray areas' of ethical research practice. Somewhat prone to guilt, I have found it difficult to rest easy with these dilemmas, which have made me question whether research can ever be truly ethical in an unethical world.

The purpose of this volume is to explore some of these gray areas and to recognize ethics in fieldwork as a highly individual experience with subjective responses for which there are no straightforward answers, easy solutions, or fixed procedures. Each of the contributors recounts their personal experiences of fieldwork and describes some of the ethical challenges and dilemmas that they faced. Their accounts are ones of tension and negotiation, uncertainty and emotion. Eszter Kovàcs and Arshiya Bose describe it as 'walking ethical tightropes.'

The benefits of sharing experiences

The doctoral thesis is a piece of work that should confidently demonstrate your intellectual abilities and make the case for your acceptance into the higher echelons of academia; it is not a space for admitting mistakes, confessing self-doubts, or engaging in introspective self-critique. Yet postgraduate students are mortals and, having undertaken a major individual project, many have issues that they need to get 'off their chests': the 'real story' behind the thesis. Likewise, while negotiating ethical conundrums along the way is the researcher's respons-ibility, there is much commonality to be found in problems, challenges, dilem-mas, frustrations, and joys. Thus there is much value in the honest exchange of experiences with friends and colleagues – to recount stories, to laugh, to cry, to get angry, or to puzzle over what happened in the research process – whether in the formal contexts of supervisory meetings, departmental workshops, confer-ence sessions, or publications (e.g., Heller *et al.* 2011), or more informally with friends over coffee, through online forums or in private correspondence. Sharing provides a space for researchers to reflect on their experiences and process their thoughts, and can even be an emotional catharsis (for example, see Skinner in this volume). The exchange of experiences also offers reassurance and mutual learning. Margi Bryant reflects in this volume that she 'personally found discus-sions with other postgraduates, sometimes spontaneous, sometimes in student-organized workshops, but always with a prevailing ethos of honesty and confidentiality, far more helpful than formal "ethics training."' This book grew out of the sharing of experiences in both formal and informal settings. Many of the contributors to this volume were grateful for the opportunity to write about ethical challenges and dilemmas that had little or no place in their formal written work yet were highly significant in shaping their research.

The challenge was to compile a coherent volume of personal reflections on a topic that was so messy, complex, and contested. The book is divided into four thematic sections: ethical challenges in the field, ethical dimensions of researcher identity, ethical issues relating to research methods, and ethical dilemmas of engagement. Each contributor was asked to focus on one particular aspect of their field experience in which ethics were significant and to discuss how they negotiated problems and issues, framing their experiences in the relevant liter-ature, providing thoughts for other researchers, and recommending helpful read-ings. Although the examples in this book are mostly drawn from fieldwork in the Global South, the content, discussions, and reflections are relevant to field research in many other countries and contexts.

Part I examines some of the broad challenges relating to doing fieldwork overseas. It starts with a wonderfully honest account by Riamsara Kuyakanon Knapp of some of the issues faced before departure, including reflections on aca-demic knowledge generation and what actually constitutes 'the field' as well as who has the power to decide about fieldwork locations. While Riam headed off to Bhutan, purportedly the happiest country in the world, Julia Tomei chose to go to Guatemala, the murder capital of the world. She discusses her choice of fieldwork location, the responsibilities that came with that choice, and its impact on herself and others. Andrew Brooks also reflects on risks and danger in the field in relation to his fieldwork in several countries of Southern Africa, which investigated controversial practices, corrupt people, and illegal behavior.

Many researchers conducting fieldwork in a foreign country need to use a dif-ferent language. Danielle Gent discusses some of the challenges of learning and operating in a second language during her time in Nicaragua and grapples with some of the ethical issues of translation and representation. Although Hayley

Leck conducted research in her home country of South Africa, she needed to use interpreters. She explores the significant role that interpreters and research assistants can play in shaping the research, and includes reflections by her main research assistant. Jenny Lunn and Alessandro Moscuzza's chapter also deals with involving other people in the research project: in their case the roles, expectations, and contributions of an accompanying partner. Their account of fieldwork together in India is set within examples from the literature of other researchers who have taken spouses, family members, and children on research trips.

Part II explores how a researcher's identity and resultant positionality vis-à-vis the research community influences different parts of the research process, including access to research participants and data collection. Much of the complexity of this issue revolves around the flexible nature of 'insider' and 'outsider' status. Girija Godbole reflects on her fieldwork experiences in a rural area that is located close to her home city in India, and yet that is in some respects a world away. She considers the roles of caste, age, marital status, class, and educational background, as well as the choice of how much of her own identity to reveal to or conceal from research participants. Rinita Dam and Jenny Lunn also recount experiences from fieldwork in India, comparing their concurrent experiences in the same city. Although working within a similar sector and using a similar methodology, they had considerably different responses from research participants which they attribute largely to three aspects of their identity: skin colour, language, and dress. Eszter Kovács and Arshiya Bose reflect on the ethical dilemmas of one specific aspect of their identity that shaped their fieldwork in Hungary and India respectively. Although gender was largely irrelevant to their research topics, they found themselves 'performing' their gender and sexuality in order to facilitate their research. Luke Taylor explores his part-insider status in Jamaica due to the presence of his family members, describing how they were invaluable in providing safety advice, facilitating access, and developing rapport, but he also reflects on some of the ethical dilemmas of involving family members in fieldwork.

Part III considers some ethical issues relating to different research methods. Margi Bryant explores the delicate position that she was in when conducting an ethnographic study in Kenya among international volunteers and local staff, and some of the ethical issues that she had to grapple with. Thomas Smith reflects on his experiences of conducting interviews in Tanzania and the role of personal emotions, feelings, and personality in shaping the process. Nora Fagerholm's chapter describes her fieldwork using participatory GIS in Zanzibar and how she negotiated ethical concerns relating to various aspects of this technique, ranging from sampling to dissemination. Virginie Le Masson explains the gendered methodology that she used for her fieldwork in India and its impact on the data collection process, while also reflecting on some of the ethical dilemmas of delegating half of the research project to a male research assistant.

Part IV explores various ethical dilemmas relating to the ways in which researchers engage with those around them, including research participants, research communities, partner organizations, and other stakeholders. Chloe Skinner reflects on the difficulties of implementing the ethical practices of informed consent, confidentiality, and anonymity in the context of her fieldwork with perpetrators and victims of violence in Israel and Palestine, as she battled to balance institutional requirements with sensitivity to the situation. Caroline Day also explores working with vulnerable people, in her case children and young people in Zambia, and describes the additional ethical procedures used to safeguard them before, during, and after fieldwork.

The next three chapters look to the other end of the social spectrum at research among elites. Kamakshi Perera-Mubarak's account of interviewing politicians and local officials in Sri Lanka includes difficulties of accessing relatively powerful people, responding to corruption, and maintaining confidentiality when the respondent is well known. Deepta Chopra recounts her frustrations when dealing with a senior civil servant in India and the ethical dilemmas surrounding her rights to access official documents, as well as her dual identity as an academic and activist. Yue Wang describes her experiences with business people in her native China, focusing on the subtleties of power play and social obligations which had an added layer of complexity as many of her interviews were conducted in restaurants over dinner. In the next chapter Stephen Jones reflects on his relationship with a collaborative partner, an NGO, during his fieldwork in Mali and the difficult position of being simultaneously an independent researcher and a member of staff, as well as the challenge of making research relevant to academia, policy-makers, and practitioners.

The final chapter looks at 'giving back,' something that all researchers need to consider. Samantha Staddon reflects on her relationship with the communities in which she carried out research in Nepal and the difficulties of deciding what to give back, whom to give back to, and when to give back. Katie Willis, co-author of *Postgraduate fieldwork in developing areas: A rough guide* (Robson and Willis 1994, 2nd edn 1997), which has been an invaluable reference for students ever since it was published 20 years ago, concludes the book with a thoughtful Afterword framing research as both a social process and an ongoing process.

The sheer range of experiences collected in this book is awe-inspiring. They span fieldwork experiences in Africa, Asia, Latin America, the Middle East, and Eastern Europe. They include research conducted in remote forests and megacities, with slum dwellers and government ministers, in desert villages and air-conditioned restaurants, with vulnerable children and business executives. The diversity of the contributors and their academic, professional, and personal backgrounds is also an interesting reflection on the varied routes from which people approach development studies. The short biographies provided at the start of this book show how the contributors have progressed from different academic disciplines and where they hope their postgraduate studies will lead in the future. This book should have wide appeal among those conducting research in the Global South, whether inside or outside academia, but particularly for undergraduates, postgraduates, and early-career researchers. The chapters are all self-standing, so the book is ideal for dipping into but equally can be read thematically or in the order presented. The Glossary will also be of help to those new to development studies, as it covers terms found in the chapters relating to development, research methodology, and ethics. While all field research is unique and individualized, there is great value in learning from the experiences of others. This book represents an exciting new resource for reference before, during, and after fieldwork and complements other publications on research methods and overseas fieldwork.

References

DARG (Developing Areas Research Group) (2009) *DARG Ethical Guidelines*, available at: www.devgeorg.org.uk/?page_id=799 (accessed 18 August 2013).

Heller, E., Christensen, J., Long, L., Mackenzie, C. A., Osano, P. M., Ricker, B., Kagan, E. and Turner, S. (2011) 'Dear Diary: Early career geographers collectively reflect on their qualitative field research experiences', *Journal of Geography in Higher Education*, 35 (1): 67–83.

Robson, E. and Willis, K. (eds) (1997) *Postgraduate fieldwork in developing areas: A rough guide*, 2nd edn, Developing Areas Research Group Monograph No. 9.

Sultana, F. (2007) 'Reflexivity, positionality and participatory ethics: Negotiating fieldwork dilemmas in international research', *ACME*, 6 (3): 374–385.

'The Nuremberg Code (1947)' (1996) *British Medical Journal*, 7070 (313): 1448.

Part I

Ethical challenges in the field

2 When does 'fieldwork' begin? Negotiating pre-field ethical challenges

Experiences in 'the academy' and planning fieldwork in Bhutan

Riamsara Kuyakanon Knapp

Actually (already) in the field

I worked, lived, and travelled in different countries in the decade before I began a postgraduate degree in human geography, a degree that required fieldwork, going to 'the field,' and being 'in the field.' As someone who had experienced different cultures, I found it unsettling that 'the field' was often conceived and spoken of as some place 'out there,' as if my venerable English university and we its denizens were a normative state that was left behind when one went 'to the field.' Like some postcolonial and feminist geographers (e.g., Staeheli and Lawson 1994; Raghuram and Madge 2006), I had problems with this 'othering' of the field, and the attendant assumption that our academic environs were the norm (Rogers and Swadener 1999). If 'field' could at its simplest be defined as 'the sphere of direct or practical participation in work or research' (*Oxford English Dictionary*), then my university and department were fields as much as anywhere else, though I did not formally focus on them as field sites. That our academic environs were not norms to be taken for granted was underscored when access to them was restricted, for example for those of us who had to apply for and successfully negotiate the visa process with the UK Border Agency in order to study in the UK.

Being 'in the field' in my department, navigating academia, was as engaging and challenging as being elsewhere, including my 'official' year of fieldwork spent in the Eastern Himalaya. This challenge was initially frustrating, as I faced the daily reality of an institution that was as socially constructed and culturally situated as any other, replete with its own contradictions, hierarchies, inequalities, and opportunities. We were taught to look critically at these structures elsewhere, presumably when we were 'in the field,' but as Driver (2000) observed, the lens was seldom turned back on ourselves within our academic setting. I found there could be large discrepancies between what was published, drawing from field research, and what was practiced in academic environs. If the 'aim of situating academic knowledge is to produce non-overgeneralizing knowledges that learn from other kinds of knowledges' (Rose 1997: 315), while this aim may have been practiced 'in the field' it was seldom brought back to 'the academy.' I observed that experiential or 'outsider' knowledge seemed little valued or understood unless couched in academic language. In academic-speak, I experienced the situation of the 'subaltern' who could not speak, 'so imbued must s/he be with the words, phrases and cadences of "Western thought" in order to be heard' (Jazeel and McFarlane 2007: 784). The irony, of course, was that at the same time my discipline of geography claimed to champion 'subalterns.'

A function of our inimitable department was to set (as well as to question!) norms for geographical enquiry in a milieu of academic ferment characterized by both dissension and cooperation. For new postgraduates, what was most evident was the cultural divide between the established 'academics' at home in the institution with a certain kind of culture, and us. Even between native English-language speakers, our different norms and subjectivities were underlain by misunderstandings. In our office there was a translation sheet pinned to the board. It was divided into three columns under the headings 'what the British say,' 'what the British mean,' and 'what others understand.' My favorite, because I heard it in seminars, was 'very interesting.' According to this sheet, when 'the British' said 'very interesting,' what they meant was 'That is clearly nonsense'! While the chart represented a gross generalization, it was funny and helped me to better understand the communication of some of 'the natives.' For many of us non-natives, our department was a field in its own right, and in our first year there, we spoke with relish of an archaeology à la Foucault (1972) and deconstruction of what went on within the walls.

As a new initiate, I had first-hand observation of the 'hegemony that the academy holds over its members, especially students and others who feel disadvantaged within the academy (due to their class, race, gender backgrounds, perhaps?) to "speak" in a certain way' (Raghuram and Madge 2006: 276). I was able to experience my department as a 'place' where you could speak your voice and they would not understand you until you became one of them. I suspect that some of us, especially those who left after a master's degree, were never understood at all. Time and further study improved my ability to communicate in language that the institution's more permanent denizens could better understand, and my sense of belonging grew. While the personal frustration at not being understood by those who through their work spoke for the 'voiceless' dissolved, the irony of the observation remained. I also realized that I had held an idealized picture of 'the academy' and those within it. The challenges (described below) that I faced navigating within 'the academy' during my 'pre-field' period tested some of the skills, subjectivities and compromises that I would also use during fieldwork. I found it helpful to situate myself as being already, in a sense, always 'in the field,' whether it be one, several, or many, distinct, diffuse, or interpenetrating.

Borrowing the shoulders of unknown giants[1]: in-text citations as an ethical challenge

> The only kinds of knowledge that are taken seriously by the Euro-American academy are those that conform to its own particular formats of writing, citation, and history.
>
> (Jazeel and McFarlane 2007: 786)

As a newly enrolled graduate student already 'in the field' in my institution, I grappled with two challenges in my early stages of producing the kind of knowledge that would be 'taken seriously by the Euro-American academy' (citation above). The first was the issue of in-text citation as a writing convention. The second was a less tangible and more perplexing issue: field site selection. While the challenges were quite different, they were alike in that they were issues generally overlooked and little written about, and my experience had not equipped me to deal with them.

It may seem ludicrous that in-text citations should have presented me with an ethical challenge, but they did. Put simply, the challenge was that I understood that we cite works in order to support our argument and show its genealogy. Thus, I felt that to make an in-text citation without actually having read the work and obtained first-hand knowledge was to make a false claim about my expertise. As naive as this line of reasoning may appear, it was bolstered by the kinds of answers I received from published academics, and the lack of advice available on the issue in academic writing guidelines (Rose 1996). There were no courses on academic writing conventions; the closest was a seminar on conducting literature reviews, where this problem does not arise since de facto a literature review is a direct engagement with the material and requires that you have read it well.

I asked my cohort how they were handling the issue when writing essays, and it was with pragmatism. You noted who established authors cited for certain ideas or statements or theories, and you cited them too, whether or not you had read the work. Searching abstracts was another method – if you made a statement without a specific reference, then the thing to do was to find papers that could be used as citations. A friend in the humanities recounted his use of 'ibid.' as the silver bullet when time was short and support was needed. An Ivy League classmate gave her handy rule of thumb, which was along the lines of 'three citations after each sentence,' though she sometimes went overboard and inserted more. Hunting for papers to use as citations was an activity in itself, which seemed to me misguided and backward. Overall, I found that little scruple existed in misrepresenting one's own scope of knowledge, as it was believed that the cost of submitting a paper with a human-capacity references list could be a lower mark or a comment on the necessity for wider reading.

Did I feel that some of the above citation practices were shady? Yes. What did I end up doing? I consulted with my supervisor, who perceptively recast the issue and nuanced my black-and-white dilemma: citing meant you were acquainted with the argument of the work – ideally through having read it. So there were shades of gray, within which I could operate with less misgiving. I made what I felt was an awkward but necessary compromise: I would seek out the work, and if it was not accessible to me – meaning I could not hold it in my hands in the case of a library book, or get a confident understanding of a key argument from an article – then I would not use it. I felt that I should always be able to answer the question 'Why did you choose to cite this work?' in a meaningful way, and not along the lines of 'because that is the standard everyone or "x" authority cites (and I copied them).' However, a circular argument remains on this issue: how can you understand an author's work thoroughly if you have not read it thoroughly? Within a system that does not accept non-pedigreed knowledge, this is a practice, I thought, that forces you to dissemble.

With more experience in reading and deeper exposure to some fields of enquiry, the reward of trying to adhere to only citing what I have read is that in my own area of expertise, I am able to gauge the substance and direction of an author's knowledge and thought through their in-text citations. Now, when I search for a paper to cite, it is to explore the literature, or because I know that the work is already out there, and has influenced my thinking beyond common sense. This issue highlighted the contextuality of institutional ethics to me. To navigate it, I needed an understanding of the culture and common practices and expectations around the use of in-text citations, which was surprisingly difficult to obtain. I moved from being very black-and-white (and in consequence quite tortured about an apparently trivial issue) to accepting the different shades of gray, being pragmatic without forgetting my ideal, and working my way through it.

The change in my understanding also had to do with my movement from a taught master's degree to a doctorate program where I was differently situated within the academic culture with the passing of time. Accordingly, the reasoning and strategies for citing also change – from using in-text citations as an indicator that you are aware of key texts and ideas to signalling your stance and allegiances as well as to attribute authorship and participate in the academic economy which trades in citations. There are implications to this. As Shirley Rose argues, scholarly citation practices are presented in 'terms limited to a view of ideas as intellectual property and of scholarly productivity as a factor in a capitalistic economy,' where 'words and ideas are widely regarded as property' (1996: 35) and language and thought are treated 'as a product of individual labor' (1996: 37). In my experience, I found that effectively, a certain kind of knowledge and knowledge production is privileged above others, and we are representing familiarity until we really are familiar. I remain troubled by the sleight-of-hand this may involve, for reasons beyond the obvious. Let me explain.

While reading around some of the ideas I have written about in this paper, I was struck to recognize Cindi Katz's phrase 'we are always already in the field,' as it was nearly the exact expression I had used above prior to seeing her work. For me, raised between cultures, and within a culture dominated by Buddhist ontology and epistemology, ideas of relativities, subjectivities, and positionalities that the social sciences engage with are pervasive. Katz's sentence begins 'Under contemporary conditions of globalization and post-positivist thought in the social sciences, we are always already in the field' (1994: 67). While we are in agreement on this view about always being in the field, I would presume to say that our understandings that have led to a shared view are different. Citing Katz for this phrase would be misleading, because I arrived at it through an entirely different reasoning process from hers, because I do not mean it as she does, because citing such a phrase would be equal to crediting authorship for what is considered a universal truth that has been engaged with at least throughout the history of Buddhist thought (the origins of which are conventionally dated to some 2,555 years ago).

Interestingly, in Katz's paper, when she first uses the phrase 'always already in the field,' there is no attribution, but she later repeats the phrase and references Koptiuch's conference paper (Katz 1994: 70). Upon further investigation into the phrase 'always already' I see that it has associations with at least Kant, Marx, Heidegger, Blanchot, Derrida, and Spivak. Could Katz be alluding to some or all of these illustrious predecessors? Perhaps, but without more familiarity, the allusion is lost on me. If I chose to cite Katz, I would be choosing to silence my experience in favor of an academic aspiration. I would be agreeing to the trademark placed by 'post-positivist thought' on concepts of relativity and subjectivity that Buddhist and other philosophies and cultures have engaged with for millennia.

Do I choose my field site or does it choose me?

My doctoral research examined the relationship between environmental conservation and Buddhist culture, and the range of field sites I could have chosen was surprisingly extensive: basically anywhere a Buddhist community existed. The geographically more accessible ones for me were in Scotland while the culturally and linguistically more accessible ones were in Thailand, with the advantages and difficulties of working 'at home.' My first choice, however, was a small nation in the eastern Himalaya. While the academic and personal reasons to propose conducting fieldwork in Bhutan were very strong, I also felt it was important that my access be granted, rather than assumed.

During course work, lectures, and seminars in our department and university, I had been exposed to ideas in critical theory and particularly postcolonial theory that appealed to me. I could see how they challenged the status quo 'out there.' At the same time, it was striking how very colonial and paternalistic and hierarchical our department could sometimes be in its operations and as a social environment. This was reflected in aspects from staff ratio and composition to mores and manners to funding structures. For example, it was noticeably easier to get funding to go and study in a former colony than elsewhere, or if you were a Commonwealth citizen. Some of these characteristics were perhaps a combination of more traditional English and 'Oxbridge' culture than endogenous to the department, itself a bastion of liberalism and anarchy compared to some of the colleges. (Lest I misrepresent my department, it should be noted that the traits I have highlighted here are but one aspect of an institution recognized by its own members as complicated and sometimes contradictory, wondrous and rewarding on the whole.)

What struck me was that in selecting field sites, it too often seemed assumed that, if logistics worked out, you could just go and do fieldwork wherever you pleased, in a globalized, borderless, and accessible world. Governments could be an obstacle, but you might get around them if their politics were 'inconvenient' to your research agenda (Crittenden 1988). I felt that this was symptomatic of a 'Western' or Euro-American hegemonic discipline that 'too easily treats problems of cultural translation and cultural difference primarily as problems of writing' and not of methodology (Sidaway 1992: 404). Notwithstanding (or perhaps *because of)* the engagements with positionality and 'giving back,' our whole process was so unselfconsciously privileged that I was disturbed. In such an environment, one in which I was undoubtedly also a beneficiary, I did not want unmindfully to perpetuate this position of privilege. One of the reasons I proposed to conduct fieldwork in Bhutan was that I would have to apply for permission to work there. The ultimate decision on whether or not I would conduct fieldwork there was not mine. It was a privilege that I hoped would be granted to me. Such a situation – where my proposed field site could reject me from the outset – powerfully demonstrated the agency of 'the field.'

Bhutan's high-value/low-volume tourism policy saved me the uncomfortable freedom facing researchers who have options on how to enter the countries where their field sites are located. At $250 a day in 2012 it would have been unaffordable as well as unethical to enter Bhutan on a tourist visa to conduct research. While I have yet to encounter the dilemma personally, this is a real issue with many ethical nuances that I feel ought to be considered prior to field site selection. As a critic (and victim and perpetrator) of rampant materialism and consumerism – consumerism of cultures too – would it sit easily with me to gain entry to a place via the tourist dollar and under false pretences? And was this actually a personal decision? As a person, I represent many things, and as a doctoral candidate, I also represented my university and the wider research community.

One result of my decision to apply for permission to conduct fieldwork in Bhutan was that I had to be prepared for well over a year of committed work without knowing whether or not I would actually be able to pursue my research in the planned context. The uncertainty of it kept me humble, optimistic, and flexible, qualities which were also useful once I was 'in the field.' Having to plan my research around being granted access also meant that I had to face issues of positionality, representation, and accountability early on, in how I approached and framed my research topic. This meant that I (if somewhat inadvertently) took the approach of the responsible social scientist who addresses the issue of research politics 'from the point wherein research is conceived' (Jazeel and McFarlane 2007: 784). In my case, it was easy to understand that 'committing to research in a place means realizing the inalienable connections between (your) theory and (their) politics' and 'the always already "public-ness" of academic knowledge production' (ibid.: 783).

A butterfly flaps its wings in Brazil[2]: external 'ethical events'

I have written about several ethical challenges that I faced while operating in the 'field' defined by the confines of my institution and department. At the same

time, some significant 'ethical events' occurred outside which affected my proposed work. One I was aware of, the other I was not. Both were similar in that
they were unforeseen yet interconnected events that affected my access to 'the
field' and subsequent positionality.

It was nearing the end of my first year as a doctoral student and the beginning
of autumn in Cambridge. The BBC was airing the first part of a three-part
program, *Lost Land of the Tiger*. I had eagerly awaited this, as the documentary
was about an expedition to camera-trap tigers in Bhutan. It was not only relevant
to my research, but also appealed to my wider interests in wildlife documentary
film. As the first episode started and the minutes passed, I saw how the story was
playing out: I grew increasingly disturbed by the portrayal of Bhutan and of the
people in front of the camera. The plot was unfolding like an Indiana Jones
movie. For me, as a fan of wildlife films and the Himalayan landscape, watching
the show was fantastic fun. For me, as a social scientist in training, who had
worked in a conservation organization and was from a tiger-range country, it
became mortifying to watch. The ignorance and misrepresentation of the local
context was painful to see. Unfortunately it was not surprising.

There is a large body of literature that critiques the practices of conservation
and conservation science that do not take social and cultural issues into account,
and these critiques exist for a reason. What was novel (for me) was seeing
reputed scientists pandering to the camera and television audience. The storyline
was clear – brave white male explorers, loaded with state-of-the-art equipment,
were going into an unknown country to discover whether tigers existed there,
and at what altitudes. The natives were friendly but not to be trusted as having
any real knowledge of their environment, and certainly not to know whether
tigers or snow leopards really existed there. Only the expert explorer scientists
and equally cool camera guys could discover this and get *real* evidence. This
was important, because the experts had theorized that Bhutan was the 'missing
link' (conveniently overlooking existing research in the country) in the tiger-
range countries, and finding tigers here would mean the possibility of saving all
tigers on the planet!

A BBC article headlined 'Lost tiger population discovered in the Bhutan
mountains' (Walker 2010) was released the day before the first episode of *Lost
Land of the Tiger* went on air. In short, the BBC claimed the 'discovery' of tigers
at high altitude in Bhutan, where their existence was already known. This may
not have made many waves outside Bhutan, but within the country feelings ran
high. There was outcry in the national paper (c.f. Pelden 2010, Phuntsho 2010)
and in online forums. Many people who had come into contact with the produc-
ers and team involved in the project felt that they had been manipulated. Evidence was given of data already existent in the country.

Of course the event is far more complex, and the story can be told from many
sides (few, alas, that would entirely exonerate the *Lost Land* team). The strong
feelings that the incident raised had antecedents in past trespasses, and already
existed to a degree in some parts of the national psyche, with justifiable reason.
The colonial history of research, extraction, and appropriation is not one that can

be ignored; the work of the BBC's *Lost Land of the Tiger* program and those involved in it reflected poorly on the international wildlife documentary and collaborating research communities, and did a disservice to the reputations of both.

How did this affect me as a researcher, an invisible part of the web of interconnectedness? Well, my research proposal was up for review with government agencies and institutions responsible for conservation research in Bhutan just after this happened. To my best ability, I made the case for conducting research responsibly, referring indirectly to the incident. The longer I waited to hear from the various organizations and people I had contacted, the more I wondered whether and to what degree the *Lost Land* incident was affecting reception of my research proposal. While my proposal was ultimately approved and permission granted for my research, this period of waiting taught me about accepting uncertainty, having an aspiration and pursuing it to one's best ability, but without clinging to it. As a result of the *Lost Land* incident, I highlighted ethics in my research approach.

At my host institution in Bhutan, one of the first things I did was to present my research plan for feedback and collaboration. 'Ethics' headed the third slide in my presentation. I wanted to clear the air and consult with my colleagues as well as let them know my aims and intentions. What came up in the cleared air after the meeting was that a more localized (un)'ethical event' had occurred. One of my Bhutanese counterparts told me this story. The summer before, an undergraduate student from a top American university had come to them by unorthodox ways, literally turning up on their doorstep in central Bhutan, and they had hosted and helped the student in her research. After she left, they did not hear from her again as they had expected, despite sending repeated emails. I was given to understand that this experience was behind the initial unresponsiveness toward my proposed research. While the story no doubt has other sides, hearing it made me realize how very connected we all were, not just at some abstract and theoretical level, but even at a very superficial level – because I had met that same student at a conference held in Bhutan the year before, which she had attended thanks to high-level contacts. She was clearly going to try to stay longer than the conference permitted to conduct field research. I had been bemused by her presence, but also impressed by her fortitude in forging on, albeit with misgivings about her approach. I did not know at the time that the institute she was going to assail was the one that would initially close their doors to me, still smarting from their experience with her and similar researchers (I also learned of other cases). (See also Staddon in this volume on following in the footsteps of other researchers and their reputation.)

Final thoughts

I have written about some of the 'ethical experiences' that I had prior to beginning fieldwork and while in the preparatory stages, and it may be that my recollections and reflections will help other researchers to consider these issues. What connects the challenges of navigating academia, citation practices, field site choice, and research ethics in the field? I would propose that the challenges above are connected by our ideas of collectivity, our illusions of individuality, and the importance of intention, speech, and action in our daily lives, wherever we are.

Upon further study and reading, I would conclude that my challenge with in-text citations was neither a ludicrous nor a trivial issue. How we cite at the very least brings up questions on the evolving nature of knowledge, on what is considered property, and on the importance of culture to these perceptions (Pennycook 1996, East 2005). How we cite has to do with how we theorize and practice theory, and how we contribute (or not) to the hegemony of 'the academy' (Raghuram and Madge 2006) and its underlying values of individuality and ownership (Hunter 1997).

Building on Rose's point about the capitalistic economy of academic citations (1996), what is the relationship of such an economy of thought and language to conducting fieldwork? In problematizing where 'the field' was and critically looking at some of the practices in our academic environs, I found that until I learned the language, non-academic life-experience was not recognized as acceptable currency. This was an engaging and valuable experience for me. In the subsequent experience of conducting fieldwork, I felt that going 'to the field' was also a way to re-legitimize knowledge, to test it through academic lenses, in order to become, if you will, a member of the academic guild and able to trade in my discipline.

While a PhD may be an intensely personal project, as students 'in the field' we also represent the research community, our institutions, our nationalities, our genders – all sorts of things, to all sorts of people. We can tramp through the world preoccupied with seeking truth 'our own' way, or try to step back to a broader view which acknowledges our interconnectedness, and thinks of those who went before us, those who are here with us, and those who will come after us, realizing the impossibility of pleasing all, and that we will be unable to fully know the implications and effects of our research (Rose 1997).

For me, cultivating a state of 'being in the field' can be conceived of as a kind of awakening to not only cultural, social, institutional, and personal subjectivities but also to human interconnectedness. It requires a physical or mental departure from daily norms – and a conscious return. One function of this is that it has given me a lens and a mirror to better understand myself and those around me, whether in my department or in a 'field site' – or anywhere, really. There will be miscommunication, offence may be taken when it was not meant, and I may be in turn 'perpetrator,' 'bystander,' or 'victim.' Developing a sense of humor and working on compassion and non-attachment have been invaluable practices.

The stories I have told here are greatly simplified accounts from one perspective only, on complex issues and events. If I have offended anyone, I extend my sincerest apologies. I would also like to thank all whom I have come in contact with and who have taught me, not least my supervisor, who is inspirational as a human being and as an academic.

Acknowledgments

In addition to family, friends, colleagues, and numerous others met along the way, I would like to express my sincere gratitude to the Department of Geography, University of Cambridge, for providing me with a challenging, engaging, and rewarding atmosphere within which to grow, and to the Royal Government of Bhutan, Ministry of Agriculture and Forests, for kindly hosting me and enabling my field research.

Recommended reading

Crittenden, R. (1988) 'Breaking the rules: Geography goes "gung-ho"', *Area*, 20 (4): 372–373.

This 'Comments' piece is part of a discussion in *Area* (1986–1988) on whether geographers should conduct research in 'Third World' countries when they cannot obtain a visa. It brings up points and asks questions that are still relevant, such as what is considered 'useful' research, and to whom.

The Professional Geographer (1994) 'Methods and techniques' section, 46 (1): 54–102.

The 'Methods and techniques' section of this volume contains key articles on critical feminist methodologies and perspectives, addressing questions of reflexivity, positionality, and politics.

Rogers, L. J. and Swadener, B. B. (1999) 'Reframing the "Field"', *Anthropology and Education Quarterly*, 30 (4): 436–440.

This piece contests the 'embedded centrality of the university' (p. 436) in relation to 'the field' and sees the academy as 'an archaic colonial position' (p. 437). It advocates an alternative framing of the university and the field, and suggests how researchers might navigate this in practice.

Rose, S. K. (1996) 'What's love got to do with it? Scholarly citation practices as courtship rituals', *Journal of Language and Learning Across the Disciplines*, 1 (3).

This elaboration and deconstruction of academic citation practice is directed at teachers but is also useful for new graduate students. Rose situates citation practices within a capitalistic economy while arguing from a Burkean (1950) perspective, and discusses some conventions of citation, giving examples of pitfalls and failings in inexperienced academic writing.

Sidaway, J. D. (1992) 'In other worlds: On the politics of research by "First World" geographers in the "Third World"', *Area*, 24 (4): 403–408.

This discusses the politics of research and representation, and the continued unequal relations between 'First World' geographers and the scholars and societies in 'Third World' countries they study. Like Crittenden's piece, Sidaway's remains relevant. He situates research within the social world that produces it, raises questions on its uses in the 'Third World' by foreign researchers, and suggests some basic working principles.

Notes

1 This refers to the metaphor 'we are as if dwarfs standing on the shoulders of giants' (*nos esse quasi nanos gigantium humeris insidentes*), which is often treated as a saying in the public domain but was first attributed to Bernard of Chartres in the twelfth century by John of Salisbury, and more famously attributed to Isaac Newton in the seventeenth century ('Standing on the shoulders of giants' 2012, Para. 2). It is most visible on the homepage of Google Scholar as 'Stand on the shoulders of giants', where it is unattributed (Google Scholar 2012).
2 This refers to the title of Edward Lorenz's talk on the 'butterfly effect' given at the American Association for the Advancement of Science in 1972 (Lorenz 1993).

References

Burke, K. (1950) *A Rhetoric of Motives*, 1969 edn, Berkeley, CA: University of California Press.

Crittenden, R. (1988) 'Breaking the rules: Geography goes "gung-ho",' *Area*, 20 (4): 372–373.

Driver, F. (2000) 'Modern geographical thought,' *Geographical Journal*, 166: 87–88.

East, J. (2005) 'Proper acknowledgment?', *Journal of University Teaching and Learning Practice*, 2 (3): 1–11.

Foucault, M. (1972) *Archaeology of Knowledge*, 2002 edn, London: Routledge.

Google Scholar (2012), available at: http://scholar.google.co.uk (accessed 10 October 2012).

Hunter, J. (1997) 'Confessions of an Academic Honesty Lady,' Grinnell College, available at: www.grinnell.edu/files/downloads/con_hj.pdf (accessed 10 October 2012).

Jazeel, T. and McFarlane, C. (2007) 'Responsible learning: Cultures of knowledge production and the North–South Divide,' *Antipode*, 39 (5): 781–789.

Katz, C. (1994) 'Playing the field: Questions of fieldwork in geography,' *The Professional Geographer*, 46 (1): 67–72.

Lorenz, E. (1993) 'Predictability: Does the flap of a butterfly's wings in Brazil set off a tornado in Texas?,' paper given at the American Association for the Advancement of Science, 1972.

Oxford English Dictionary, available at: www.oed.com/view/Entry/69922£eid190158717 (accessed June 2011).

Pelden, S. (2010) 'Not just here but higher,' *Kuensel*, 25 September, available at: www.kuenselonline.com/2011/?p=15568 (accessed 10 October 2012).

Pennycook, A. (1996) 'Borrowing others' words: Text, ownership, memory, and plagiarism,' *TESOL Quarterly*, 30 (2): 201–230.

Phuntsho, K. (2010) 'Stealing our thunder,' *Kuensel*, 25 September, available at: www.kuenselonline.com/2011/?p=15567 (accessed 10 October 2012).

Raghuram, P. and Madge, C. (2006) 'Towards a method for postcolonial development geography? Possibilities and challenges,' *Singapore Journal of Tropical Geography*, 27 (3): 270–288.

Rogers, L. J. and Swadener, B. B. (1999) 'Reframing the "field",' *Anthropology and Education Quarterly*, 30 (4): 436–440.

Rose, G. (1997) 'Situating knowledges: Positionality, reflexivities and other tactics,' *Progress in Human Geography*, 21 (3): 305–320.

Rose, S. K. (1996) 'What's love got to do with it? Scholarly citation practices as courtship rituals,' *Journal of Language and Learning Across the Disciplines*, 1 (3), available at: http://wac.colostate.edu/llad.issues.htm.

Sidaway, J. D. (1992) 'In other worlds: On the politics of research by "First World" geographers in the "Third World",' *Area*, 24 (4): 403–408.

Staeheli, L. A. and Lawson, V. A. (1994) 'A discussion of "women in the field": The politics of feminist fieldwork,' *The Professional Geographer*, 46 (1): 96–102.

'Standing on the shoulders of giants' (n.d.) *Wikipedia*, available at: http://en.wikipedia.org/   w/index.php?title=Standing_on_the_shoulders_of_giants&oldid=518009957 (accessed 10 October 2012).

Walker, M. (2010) 'Lost tiger population discovered in Bhutan mountains,' BBC News, September 20, available at: http://news.bbc.co.uk/earth/hi/earth_news/newsid_8998000/8998042.stm (accessed 13 October 2010).

3 'I always carried a machete when travelling on the bus': ethical considerations when conducting fieldwork in dangerous places

Experiences from fieldwork on the sustainability of biofuels in Guatemala

Julia Tomei

Guatemala is one of the world's most dangerous countries, with some 6,500 murders in 2009. In recent years, Guatemala has been crippled by soaring levels of violent crime and impunity.... Citizens from all walks of life increasingly face the threat of murder, kidnapping, extortion, gang violence and shootouts between rival drug trafficking organizations.

(International Crisis Group 2010)

You might wonder why on earth I chose to do my doctoral fieldwork in Guatemala, particularly given that my research on the sustainability of biofuels could have been studied virtually anywhere in the world. Was I mad, courageous, or simply naive? In this chapter I give some personal reflections on my experiences of negotiating multiple dangers before, during, and after fieldwork. My perception of danger – principally physical and emotional – led me to consider carefully my security in the field, whether or not I wanted to return to Guatemala for a second field season, and also how my research design could be altered to mitigate risk. I hope that the challenges I faced and the ways in which I dealt with them may be of some use to other researchers considering embarking on fieldwork in risky settings.

Going it alone

For many postgraduate researchers fieldwork is an integral element of their degrees. The idea of the first field experience as a 'rite of passage,' in which the fieldworker must 'go it alone' relatively unaided, is a pervasive one (Lee 1995). The dangers that may come with this experience – perhaps from research undertaken with risky groups or in risky settings – can be regarded as 'part and parcel' of acceptance by the academy. For many of us, danger permeates our research and may exaggerate an already intense research experience abroad. Yet there are surprisingly few peer-reviewed, formalized accounts of dangers to fieldworkers (Lee 1995, Belousov *et al.* 2007).

Institutional guidelines on ethics generally follow the medical injunction to 'do no harm' to participants (Patai 1991: 137). Harm to the researcher, however, largely falls under institutional 'risk assessments.' Since such assessments must be applicable to different types of research and a range of academic disciplines, they are inherently general and often fail to identify or help to mitigate the risks to researchers in the field. Typically, discussions of danger are limited to those risks faced by participants (Dickson-Swift *et al.* 2008; Dowling 2010). In this chapter I take Lee-Treweek and Linkogle's (2000: 1) understanding of danger as the 'experience of threat or risk with serious negative consequences that may affect the researcher, participants and other groups in society.'

In their collection of reflexive writings on danger in the field, Lee-Treweek and Linkogle (ibid.) identify four key areas of danger, cautioning that these are neither exhaustive nor mutually exclusive: physical, emotional, ethical, and professional dangers. This 'framework for danger' provides a useful structure for thinking through the hazards associated with carrying out field research – both to the participants and to the researcher (see also the chapters by Brooks, Skinner, and Taylor in this volume exploring specific aspects of risk to the researcher, gatekeepers, and participants).

A rational choice?

Before I began my doctoral research, I had investigated fox behavior in isolated regions of Chile, agricultural biodiversity in rural Peru, and soy-biodiesel production in Argentina. I had also travelled throughout Latin America; I spoke Spanish and had a long-standing interest in the history, culture, and politics of the continent. When I was given the opportunity to undertake doctoral research on the sustainability of biofuels, I was eager to continue my association with the region by conducting field research in Latin America. For a number of reasons Guatemala emerged as the lead contender: it had the greatest potential in Central America for the production, trade, and consumption of biofuels due to high yields of sugarcane and palm oil (USDA 2012); it was exporting ethanol to the European Union; I was offered institutional support by the public university; and it was a relatively under-researched field in an under-researched region.

Encouraged by offers of support and interest, I decided that my research would focus on the social impacts of biofuels in Guatemala and embarked on planning and designing the field research. However, as I started my planning, I read a report by the International Crisis Group (see the quote at the start of this chapter), which for the first time brought home to me the volatile security situation in Guatemala.

Guatemala is a country located in one of the most violent regions of the world ('The rot spreads' 2011). The 36-year civil war was arguably the region's bloodiest conflict and claimed more than 200,000 lives (REMHI 1998). Since the end of the civil war in 1996, Guatemala has faced escalating violent crime – attributed to organized crime, drug trafficking, and gangs – which has affected every stratum of society (International Crisis Group 2010). Furthermore, it is the third most unequal country in the world in terms of income distribution; more than half the population live below the poverty line, yet Guatemala also has the highest number of helicopters per capita in the world (Lowenthal 2009, CESR 2010). The failure of successive governments to investigate human rights abuses, accusations of corruption, and the growing crime rate has perpetuated a culture of impunity within Guatemala (UNDP 2011). Added to this are environmental risks from volcanoes, earthquakes, and tropical storms, which regularly render parts of the country inaccessible.

Although Guatemalans live with these real dangers on a daily basis, when I looked at the UK Foreign and Commonwealth Office website it had no restrictions on travel to Guatemala. It did, however, advise caution in Guatemala City due to the crime rates and a rise in kidnappings, and it recommended against travel on public transport. The report by the International Crisis Group also highlighted the dangers of using public transport, of walking the streets – even during the day – and of travelling by taxi. Reading these various reports I became concerned not only about the risks to my personal safety, but also the impacts on my mental wellbeing. How, I wondered, was I, a Londoner, who was used to being independent and walking everywhere, going to be able to cope with being cooped up within four walls?

Testing the waters

Levac *et al.* (2010) suggest that one way to mitigate the unexpected during fieldwork is to carry out a scoping or pilot study before embarking on the full field project. I decided that this was a sensible idea that would enable me not only to refine the research questions but also to see for myself whether field research in Guatemala would be possible. The scoping study was an incredibly useful experience in that it enabled me to contextualize my research and to discern whether I was asking the 'right' questions (I wasn't).

However, the five-week scoping visit unfortunately did nothing to assuage my concerns regarding personal safety. This period witnessed a series of alarming events, including the kidnapping and murder of a Guatemalan social researcher who had been investigating corruption,[1] the surveillance of and threats toward a close friend and colleague, and being robbed myself at gunpoint at a set of traffic lights in Guatemala City. I had also underestimated the extent to which I resented being unable to leave unaccompanied the armed and gated *colonia* where I was staying. Despite this, I was bowled over not only by the warmth, hospitality, and interest shown toward myself and my research, but also by the resilience of Guatemalans in the face of such constraints on their day-to-day lives.

On returning to the UK, I found myself recounting my 'war stories' to family, friends, and colleagues in a light, semi-comical way. I think this was partly because others are often more interested in the extreme, the dangerous, and the different, than the more mundane day-to-day lived experiences. However, I was also wary of causing too much concern to others and, despite the apparent dangers, I did not want to be forbidden from going back. Perhaps this was selfish, but I had grown attached to Guatemala and its people. I also felt that I had reached a stage in my research that made it difficult to locate it elsewhere and, finally, wondered whether I had simply been unlucky during my first field experience and was therefore overstating the risks.

I spent the next few months reflecting on the scoping visit and contacting other researchers, mainly North American, who had recent experience of field research in Guatemala to get their thoughts and advice. Advice ranged from the incredible (e.g., 'I always carried a machete when I travelled on the bus') to the more sensible (e.g., 'always use a trusted driver'). I also contacted embassies and referred to websites, blogs, and friends known to have travelled in Central America (see also Taylor in this volume on consulting various sources of information and advice prior to travel).

The overriding message was that it was possible to do field research in Guatemala but that I should take certain measures to mitigate personal risk. It was also suggested that I might be relatively safe as a *gringa* (white foreign woman) in Guatemala (however, see also Bourgois 1990) and that as a newcomer I would be less vulnerable than Guatemalan researchers and activists. Indeed, foreign 'accompaniers' are often drafted in to the country to protect figures such as Guatemalan human rights advocates (as in the Guatemalan Solidarity Network international 'accompanier' program), demonstrating the tragic reality of the dispensability of native lives in a country with such high levels of impunity.

Back for more

As a result of this consultation and discussion with my fellow researchers, friends, and contacts I decided to proceed with Guatemala as my fieldwork location. However, I adapted the research design to assuage concerns about my safety; it moved from what would have been an emphasis on the community level using mainly ethnographic tools to one that also incorporated field research in Brussels and a greater focus on Guatemalan elites. Without going into too much detail, I feel that this evolution was beneficial in two key ways: first that it made the research more policy relevant, and second that it reduced some of the dangers I was likely to encounter while in Guatemala. This included the professional danger that I would not be able to get sufficient data should I be unable to spend extended periods in rural communities, but also the personal, physical danger.

The second phase of field research, for which I was in Guatemala for seven months, was thankfully a very different experience. A number of strategies helped me to live with the danger I perceived; these included living with a trusted Guatemalan friend in a secure location; heeding the advice of locals about what was safe to do and what was not; and never leaving home with more than I was prepared to lose. In addition, frequent communication with supervisors, friends and family enabled me to assuage their concerns, while keeping a diary allowed me to reflect on the research, as well as on issues of danger and ethics. While the threats to my personal safety were no fewer, my perceived exposure to physical danger lessened as I grew more confident over time. Peterson (2000) draws attention not only to the spatial dynamics of danger, but also its temporal contingency. While I was never physically harmed during my first fieldtrip, the fear of physical danger was almost paralyzing; this contrasts with the second fieldtrip, during which I was able to acclimatize to the realities of life in Guatemala and felt much more comfortable. In this second phase of field research, I perceived greater exposure to what Lee-Treweek and Linkogle (2000) term the emotional and ethical dangers of field work.

On the face of things, the subject of biofuels is a seemingly innocuous research topic. However, as I discovered, in Guatemala biofuels are, for some, synonymous with highly sensitive debates about land access, embedded racism, and colonialism. Although I had already spent some years researching the sustainability of biofuels, and was well versed in debates about land grabbing and loss of rural livelihoods, I was not emotionally prepared to hear personal accounts of the violent evictions that rural people experienced due to the expansion of monocultures. This led me to reconsider the possible ethical dangers faced by research participants (e.g., Ybarra 2010) and the potentially extractive nature of my research (see Sidaway 1992, England 1994). Many participants expressed their gratitude that I, a *gringa*, was taking an interest in their struggles and hoped that I would take their stories home to share with others where they might have a greater impact (see Patai 1991). However, I could not help but feel that I had gained invaluable insights into my research at the expense of participants reliving painful memories.

Danger, in addition to the usual pressures and 'blues' of fieldwork (e.g., see Scheyvens and Nowak 2003, Heller *et al.* 2011), contributed to what was an intense field experience. Toward the end of my time in Guatemala I certainly experienced a sense of fatigue due to the unrelenting pressure I placed on myself to generate data, to my dependency on others, and to what Heller *et al.* (2011: 90) call 'struggles with abstractions and things "bigger than ourselves."' Acknowledging England's statement that 'the researcher cannot neatly tuck away the personal behind the professional, because fieldwork is personal' (1994: 85), I questioned the impact that the (at times terrifying and stressful) research setting had on the data I had produced. Could my admiration of respondents, and their incredible tenacity and will in the face of extreme inequality and everyday injustices, or my heightened sensitivity and perception of danger, alter my interpretation of the field?

Heading home

Upon my return to the UK, when speaking to Guatemalan friends I was frequently exhorted to 'take care,' their perception being that London was a dangerous place. Indeed, some readers will have read this account wondering whether the dangers were overstated, while others may see commonalities with their own research experiences. So, perhaps 'risk is in the eye of the beholder' (Dickson-Swift *et al.* 2008: 133). Ultimately, it is for the individual researcher to decide what an acceptable level of danger is; if the researcher perceives an activity as risky then steps must be taken to manage those risks.

With the benefit of hindsight, I can look back and say that I do not regret my decision to go to Guatemala. After only a few weeks, I had grown to love the country with its fascinating mix of environments, cultures and worlds, and its friendly, engaging, and resilient people. The field research, although an intense, emotionally draining, and at times frightening experience, was ultimately rewarding, enlightening, and life-changing. However, I did not always feel that way. I think I was naive when I initially chose to focus on Guatemala, a decision perhaps taken lightly due to a (misplaced) confidence that came from having previously undertaken fieldwork in Latin America.

Recommendations for evaluating and managing dangers in the field

Every research setting, and the geographical, political, social, and cultural context within which the research takes place, will be different. Whether at home or abroad, the researcher may have to negotiate danger, although doing field-work in an unfamiliar cultural setting inevitably adds a layer of complexity. It is therefore difficult to provide general tips to the concerned researcher but, drawing on my own field experiences as well as the literature, I would like to suggest some steps that may be taken before and during field research. These steps revolve around planning, flexibility, and support.

First, evaluating potential danger is an essential part of the planning stage of fieldwork. Dickson-Swift et al. (2008) argue that consideration of risk has typic-ally been limited to participants, while risks to the researcher are only considered in a cursory, ad-hoc manner once in the field. Indeed this is true of most institu-tions, which will require the researcher to submit a risk assessment form in addi-tion to an ethical approval form. These are generic by nature and may exclude the non-physical dangers discussed in this chapter. In addition, while it is important to fully evaluate the dangers identified in risk assessments, others may emerge once in the field (Belousov et al. 2007); again, the psychological risks to researchers are typically overlooked by institutions (Peterson 2000; Belousov et al. 2007). The steps I undertook to evaluate the dangers to myself are not typic-ally required as part of institutional health and safety procedures, and I would advocate that the researcher looks beyond the standard questions of the typical risk assessment template. This will allow him or her to make a more informed decision about the dangers involved and how they can be mitigated. I would also recommend talking to someone in the home institution about whether insurance is provided and the terms and conditions of that provision.

Second, a flexible research design can help the researcher to expect the unex-pected, enabling him or her to respond to new situations as and when they arise without jeopardizing data collection and research validity. Pilot or scoping studies are one way for a researcher to evaluate the risks and to foreshadow issues such as ethics, representation, and health and safety (Sampson 2004, Levac et al. 2010). This may then lead to modifications to the research design, enabling the researcher to develop coping strategies while in the field, as I did.

Sluka (1990) recommends that researchers have a plan of escape to remove themselves from the field should direct physical threats arise. This may seem extreme, but it highlights the importance of reflecting on what an 'acceptable' level of risk will be (Lee 1995, Linkogle 2000).

Third, it is important to have strong personal and institutional support both at home and in the field. I see many truths in this comment by Linkogle:

> Researchers may be driven by a commitment to their discipline and a quest to generate new knowledge but this does not provide any special protection in the field. In fact it can create a perilous sense of hubris.... The romance of the lone researcher contributing to a body of scholarship through individual acumen and hard work facilitates an ethos of self-reliance which can make it difficult for researchers to ask for help ... it is also hard for institutions to grasp how to provide such assistance given a research culture that elevates individual scholarship.
>
> (Linkogle 2000: 145)

Since research can be an isolating experience, Dickson-Swift *et al.* (2008) recommend debriefing with friends, family, or colleagues. Field diaries can be a good way of dealing with the emotional and ethical dangers, as well as providing a reflexive tool which can be consulted when analysing and writing up findings (see Punch 2012).

Finally, prior to going in to the field, researchers may also find it useful to consult widely in order to find out how others have coped with field research in that particular setting. If your supervisors are unfamiliar with the research context, do not be afraid to contact others who have conducted research in your area of interest and/or research setting. Similarly, while in the field, listen to local people; if you are advised not to visit somewhere, don't go!

This chapter has outlined a range of dangers that I faced while in the field and some of the associated ethical dilemmas. It raises the question of whether it is right to put ourselves in potentially risky settings for our research, perhaps against the wishes of our family and friends. Drawing on the literature, I have also broadened the conceptualization of danger, which is typically limited to the risk of physical harm, to include the professional, emotional, and ethical dangers that research in risky contexts may present both to us and to our participants. This chapter urges researchers to go beyond the classic risk assessment template to ensure fully informed decisions when entering potentially dangerous field sites.

Recommended reading

There is perhaps a surprisingly limited literature on the danger – in all its guises – of undertaking fieldwork in the Global South. Many of the references cited in this chapter may provide useful reading on danger in the field, but I found the following two especially useful for conceptualizing about danger and the ways in which risks might be mitigated.

Lee, R. M. (1995) *Dangerous fieldwork*, Qualitative Research Methods Series Vol. 34, London: Sage University Paper.
Lee-Treweek, G. and Linkogle, S. (2000) *Danger in the field: Risk and ethics in social research*, London: Routledge.

Note

1 Emilia Quan was working for the *Centro de Estudios y Documentación de la Frontera Occidental de Guatemala* (CEDFOG), an NGO working to promote social justice in the department of Huehuetenango. Her supposed kidnappers were later lynched by the local population ('Asesinan a socióloga Emilia Quan en Huehuetenango' 2010).

References

'Asesinan a socióloga Emilia Quan en Huehuetenango' (2010) *El Periodico*, 9 December, available at: www.elperiodico.com.gt/es/20101209/pais/185470/ (accessed 9 March 2011).
Belousov, K., Horlick-Jones, T., Bloor, M., Gilinskiy, Y., Golbert, V., Kostikovsky, Y., Levi, M. and Pentsov, D. (2007) 'Any port in a storm: Fieldwork difficulties in dangerous and crisis-ridden settings', *Qualitative Research*, 7 (2): 155–175.
Bourgois, P. (1990) 'Confronting anthropological ethics: Ethnographic lessons from Central America', *Journal of Peace Research*, 27 (1): 43–54.
CESR (Center for Economic and Social Rights) (2010) *Guatemala: Making human rights accountability more graphic*, Fact Sheet No. 3, Madrid and New York: Center for Economic and Social Rights.
Dickson-Swift, V., James, E. L., Kippen, S. and Liamputtong, P. (2008) 'Risk to researchers in qualitative research on sensitive topics: Issues and strategies', *Qualitative Health Research*, 18 (1): 133–144.
Dowling, R. (2010) 'Power, subjectivity, and ethics in qualitative research', in I. Hay (ed.) *Qualitative research in human geography*, Toronto: Oxford University Press Canada.
England, K. V. L. (1994) 'Getting personal: Reflexivity, positionality and feminist research', *The Professional Geographer*, 46 (1): 80–89.
Heller, E., Christensen, J., Long, L., Mackenzie, C. A., Osano, P. M., Ricker, B., Kagan, E. and Turner, S. (2011) 'Dear Diary: Early career geographers collectively reflect on their qualitative field research experiences', *Journal of Geography in Higher Education*, 35 (1): 67–83.
International Crisis Group (2010) *Guatemala: Squeezed between crime and impunity*, Latin America Report No. 33, Brussels: International Crisis Group.
Lee, R. M. (1995) *Dangerous fieldwork*, Qualitative Research Methods Series Vol. 34, London: Sage University Paper.
Lee-Treweek, G. and Linkogle, S. (2000) *Danger in the field: Risk and ethics in social research*, London: Routledge.
Levac, D., Colquhoun, H. and O'Brien, K. (2010) 'Scoping studies: Advancing the methodology', *Implementation Science*, 5: 1–9.
Linkogle, S. (2000) '*Relajo*: Danger in a crowd', in G. Lee-Treweek and S. Linkogle (eds), *Danger in the field: Risk and ethics in social research*, London: Routledge.

Lowenthal, A. F. (2009) *Central America in 2009: Off the US radar*, Brookings Institution, Washington, available at: www.brookings.edu/opinions/2009/0106_central_america_lowenthal.aspx (accessed 9 March 2011).

Patai, D. (1991) 'US academics and Third World women: Is ethical research possible?', in S. B. Gluck and D. Patai (eds), *Women's words: The feminist practice of oral history*, London: Routledge.

Peterson, J. D. (2000) 'Shifting definitions of danger', in G. Lee-Treweek and S. Linkogle (eds), *Danger in the field: Risk and ethics in social research*, London: Routledge.

Punch, S. (2012) 'Hidden struggles of fieldwork: Exploring the role and use of field diaries', *Emotion, Space and Society*, 5: 86–93.

REMHI (Proyecto Interdiocesano de Recuperación de la Memoria Histórica) (1998) *Guatemala: Nunca Más*, Guatemala City: Oficina de Derechos Humanos del Arzobispado de Guatemala.

Sampson, H. (2004) 'Navigating the waves: The usefulness of a pilot in qualitative research', *Qualitative Research*, 4: 383–402.

Scheveyens, H. and Nowak, B. (2003) 'Personal issues', in R. Scheyvens and D. Storey (eds), *Development fieldwork: A practical guide*, London: Sage.

Sidaway, J. D. (1992) 'In other worlds: On the politics of research by "First World" geographers in the "Third World"', *Area*, 24 (4): 403–408.

Sluka, J. A. (1990) 'Participant observation in violent social contexts', *Human Organisation*, 49: 114–126.

'The rot spreads' (2011) *The Economist*, 20 January, available at: www.economist.com/node/17963313?story_id=17963313 (accessed 9 March 2011).

UNDP (United Nations Development Programme) (2011) *Guatemala: Hacia un estado para el desarrollo humano*, Informe nacional de desarrollo humano 2009/ 2010, Guatemala City: UNDP, available at: http://desarrollohumano.org.gt/indh2010_descarga (accessed 9 March 2011).

USDA (United States Department of Agriculture) (2012) *Guatemala biofuels annual: A big splash of ethanol and a drop of biodiesel*, Washington, DC: USDA, Foreign Agricultural Service.

Ybarra, M. (2010) *Living on scorched earth: The political ecology of land ownership in Guatemala's northern lowlands*, unpublished thesis, University of California Berkeley.

4 Controversial, corrupt and illegal: ethical implications of investigating difficult topics

Reflections on fieldwork in southern Africa

Andrew Brooks

> The salaries, that was too low for us to sustain our working.... It was like slavery. It was just survival of the fittest.

> I was involved in an industrial accident. The Chinese visited the hospital and made some promises to ensure my wellbeing.... Unfortunately after I was discharged it was a different story.... The management at the company contrived with the hospital to destroy or hide the file.... Without this information it is very difficult ... for me to be compensated in the future.

These quotes come from interviews I carried out with machine operators who were employed in a Chinese clothing factory in Zambia and had worked in awful conditions. The latter worker had lost his right forearm in an industrial accident and since then, in his own words, had 'become a beggar.'

As a master's degree student, I sat opposite him in a small concrete house and patiently asked my carefully prepared questions. My eye was unavoidably drawn to the stump of his arm dressed in simple bandages. Cut away below the elbow, his hand and forearm had been ripped away by textile machinery. As I interviewed him, he gestured with the limb in response to my verbal probing, alternating between calm and considered answers and expressions of outrage toward his former employer. His treatment was both morally outrageous and illegal, as he had never received the compensation payments due under Zambian law and this was illustrative of the horrific working conditions across the factory (see Brooks 2010).

For me personally, and as an early-career researcher, this fieldwork experience was deeply unsettling, but documenting these issues became part of a project which has contributed to important debates on the controversial nature of Chinese investment in Africa (see Carmody *et al.* 2011; Giese and Thiel 2012). Such moments are examples of the type of difficult, but also prescient issues, with which social research sometimes struggles to engage (Thomas 1993; Hall 2012).

In this chapter, I would like to make the case for intensive qualitative research, which uncovers and seeks to explain the causes of inequality.

Furthermore, following Paul Rabinow's thought-provoking *Reflections on Field-work in Morocco* (1977/2007), I argue that social research should be 'problem-orientated' and that the greatest social problems are found among poor people in the Global South. Frequently the most pressing problems involve difficult-to-research topics, which are challenging to engage with, especially when adhering to increasingly restrictive institutional risk assessment and ethical guidelines.

My reflections in this chapter draw on the 'messiness' of cross-cultural field-work, for which researchers cannot always be prepared (Rabinow 1977/2007). Research about inequality and illegal activity is frequently brief, ephemeral, and shallow and can fail to take into account the drivers of poverty and modes of exploitation. On the other hand, prolonged fieldwork can enable the researcher to develop lived experiences and to enhance rigor, which counteracts the shal-lowness of 'rapid' research interventions (Gladwin *et al.* 2002). Real commit-ment to field research is frequently required to enable access, build trust, and understand difficult topics, as well as to enable the researcher to make their own judgments as to what are (or are not) risky or ethical research interventions.

In this chapter, I discuss some of the ethical and practical challenges of researching difficult topics which involve controversial, corrupt, and illegal eco-nomic activities. I draw on my experiences from fieldwork in southern Africa using three case studies to illustrate my own successes and failures: the con-cealed second-hand clothes trade in Mozambique (Brooks 2012a, 2013), corrup-tion in Japanese used-car imports in South Africa and Mozambique (Brooks 2012b), and labor disputes and Chinese investment in Zambia (Brooks 2010). Through discussing my interactions with informal market workers, international traders, corrupt officials, factory workers, and Chinese business people, I explore the ethical dilemmas of dealing with people engaged in corrupt, exploitative, or illegal activities, either as perpetrators or as victims. There are various issues to contend with including access, positionality, and power dynamics. I also con-sider practical and theoretical constraints, as well as the strengths and weak-nesses of different methodological tools when dealing with these groups of people. My concluding reflections discuss power relations that I experienced in fieldwork within and beyond southern Africa.

A controversial topic: the second-hand clothes trade in Mozambique

Many used clothes donated to charities are exported from the Global North and sold in markets in Africa. This trade pattern is controversial because of the per-ceived dishonesty of charities re-selling donated clothing, the high profits made by the private companies involved, the smuggling of used-clothing imports across international boundaries, and the negative effect on local clothing indus-tries (British Heart Foundation 2011; Hansen 2000). I researched the trade in second-hand clothes from the UK to Mozambique in my doctoral thesis and linked publications (Brooks 2012a, 2012c, 2013). When reviewing the literature, I found that the trade had attracted relatively scant academic attention and, as I began to investigate this topic in greater depth, I realized it was going to pose a variety of different ethical challenges.

The commodity chain often begins with charities – including Oxfam, the Salvation Army, and YMCA – accepting donations of unwanted clothing. These organizations then sell the clothes on to companies for export to the Global South and profit from this trade. The trade in second-hand clothes has a negative impact on industrial development in Africa, as imported garments out-compete domestically produced clothes and/or eliminate the opportunity for new apparel industries to develop (Brooks and Simon 2012).

When I was planning my fieldwork, one of the models for research I considered was to work with a partner organization, as Kleine (2008) had done with wine cooperatives in Chile, or Le Mare (2007) did with Traidcraft plc. This type of research engagement can be advantageous, facilitating access to key individuals, but the capacity for critical discourse may be limited by the preferences for certain outcomes and other 'political' factors (Chataway *et al.* 2007). When planning my research, I held firm to the idea that it was *right* to make charities and the businesses they work with the subject of critical analysis and to critique their activities, as I was concerned that previous research (e.g., Baden and Barber 2005; Hansen 2000) had not fully investigated the negative impacts. From my moral standpoint, these actors and the controversial economic activities in which they were engaged should not escape rigorous scrutiny.

After deciding upon this approach, the major issue was the accessibility of potential research participants at different stages in the commodity chain. My access to different charity workers, business people, and traders was reflective of the relative agency that those groups and individuals had in trade networks. I began the research by contacting collecting organizations in the UK by email or telephone, but the responses I received were predominately refusals to participate. In general this was a very frustrating process, and when I was able to interview people there were practical challenges: repeat phone calls and emails were required to arrange and rearrange meetings as participants shifted dates and appointments (for similar experiences see chapters by Chopra, Dam and Lunn, and Perera-Mubarak in this volume).

The senior charity managers whom I interviewed were careful, cautious, and guarded in their responses to my questions. For example, when I interviewed the General Manager of Oxfam Wastesaver about the export process, I asked, 'How does a container get to, say, Lagos?' He was quick to question whether that was 'a Freudian slip' on my part; he wanted to know if I was trying to catch him out and discover if Oxfam were exporting to Lagos (when the import of used clothing to Nigeria was illegal). In actuality, the question I asked was a follow-up question and I had said Lagos as it was the first large West African port that came to mind. This example demonstrates how the General Manger was acutely aware of the responses he was giving in the interview and wanted to respond accurately and protect the image of Oxfam.

My interviews in the UK illustrated how the collection, sorting, processing, and export of second-hand clothes to developing countries are background fundraising activities which are concealed by NGOs in contrast to the overt charitable programs that are promoted in their marketing materials (Smith 2004). These organizations were aware of the potential for critical research to impact negatively upon their income-generating activities (Baden and Barber 2005; British Heart Foundation 2011).

By contrast, in Mozambique, poor market traders who sold imported second-hand clothes were less knowledgeable about the controversy that surrounded the trade. These relatively impoverished people were willing to participate in the study, but this raised ethical concerns as 'giving participants a voice risks revealing their survival strategies' (Manzo and Brightbill 2007: 33). Throughout this project I did not want to do anything that would harm the incomes of my participants, but I also believed that it was important to examine the broader impacts of second-hand clothing imports on economic development across Mozambique and sub-Saharan Africa, rather than considering the interests of the small group of traders I worked with in isolation (see Brooks and Simon 2012). Furthermore, the purpose of the research was always explained in advance; the market traders freely participated and each was provided with a *per diem* (see further discussion in Brooks 2012a).

While undertaking my 14 months of fieldwork, I continually assessed the potential risks that my research posed to the individual livelihoods of market traders and concluded that the possible negative effects were absolutely minimal. The immediate outputs of my research (a PhD thesis and linked publications) were unlikely to influence the future availability and market for second-hand clothes in Mozambique. Indeed, I have subsequently presented my research to charity and business audiences in the UK without having any impact on second-hand clothing collections, or downstream effects on the supply of used clothing to the research participants I worked with in Africa. Here there is also a frustrating contradiction and duplicity on my part. Ideally (and unrealistically), I would want my research to have an impact on economic policy (i.e., controls on second-hand clothing imports) and positively influence trading relationships between the Global North and Africa (which has not occurred). Thereafter, if this were to occur, I would find it upsetting if the poor people with whom I worked lost their livelihoods, even if the prospective change in policy were to have macro-scale benefits for African clothing industries, which I would welcome (see Brooks and Simon 2012).

One of the more practical difficulties that I faced in carrying out research in Maputo (Mozambique) was in sampling, which, as with many projects undertaken in African cities, was difficult because I had limited prior knowledge about the population structure. As Davis (2004: 24) found, there are 'formidable theoretical and empirical problems involved in studying the urban poor,' and these challenges may be especially difficult to overcome when working with people who are involved in informal-sector activities such as trading used clothes, which can be concealed from local authorities. This often means that a more reflexive approach is needed (Kapferer 1972; Zarkovich 1993). When I undertook research in Maputo with informal second-hand clothing hawkers who worked on the margins of legality, sampling had to be pragmatic and opportunistic, as research on such trading activity could have brought unwanted attention from the police for both the participants and myself (Kamete and Lindell 2010; Paasche and Sidaway 2010).

One of the topics that I had wanted to gather information on from Mozambican market traders was the up-stream processes: what was the structure of the commodity chains which connected their sales activity to second-hand clothing donations in the UK and elsewhere? Based on previous pilot research, I had anticipated that these individuals might have a lack of knowledge, but still hoped to gain some information or draw upon them to contact people who did have the answers. This proved unsuccessful.

Instead, I attempted to interview the importers in Maputo directly, but the Indian-owned import companies which bring second-hand clothes to Mozambique proved to be far more difficult to access. I first attempted to make social contact with them through expatriate networks, which previous research had shown to be a successful method for contacting difficult-to-access key informants (Hansen 2000; Thomas 1993) (see also Wang in this volume on using social networks to access informants). The Indian importers were distrustful of me; in part this can be accounted for by the fact that the South Asian expatriate community did not tend to mix with other international groups in Mozambique and had a distinct socio-history (Pitcher 2002). I later tried to interview Indian managers at their business premises. Sometimes it was possible to have a 'doorstep interview' whereby I could ask a few questions, but more frequently there was a dismissive response; I was told to return later or to go elsewhere or was just ignored (again see chapters by Chopra, Dam and Lunn, and Perera-Mubarak in this volume for similar frustrations). One importer was outwardly hostile and I had to leave rapidly. This was at times a very unenjoyable and even traumatic experience and proved to be an unassailable problem, as ultimately participation in interviews has to be voluntary according to ethical research practice (Manzo and Brightbill 2007).

In an attempt to overcome this problem of access to key informants, I experimented with other techniques. I made observations of second-hand clothing warehouses and asked workers (such as security guards and porters) questions, but they were also reluctant to be interviewed. In addition, my local research assistant visited the warehouses alone and attempted to interview staff. It was envisaged that he would attract less hostile attention than a Western researcher, but in fact he encountered similar problems. These additional informal research processes did allow me to identify principles of the trade processes and also to triangulate information acquired elsewhere. However, the activities of importers did remain to a certain extent unknown and when writing up the research in my thesis I encapsulated them by using the concept of a 'black box' (Brooks 2012c). The inputs (imports of used clothing from the Global North) and outputs (wholesale of second-hand clothes) are known, but there is a lack of clarity as to what happens between the stages, and knowledge of these power-laden spaces is policed by inaccessible research non-participants (Kitchin and Fuller 2003; Latour 1993).

Corrupt practices: used car imports in South Africa and Mozambique

In contrast to the second-hand clothing industry, I was more successful in engaging used-car importers in South Africa and Mozambique (see Brooks 2012b). This was principally because I was able to cultivate social ties with key individuals in trade networks, although in this case there were limits as to how far I could take the research because of the potential dangers and risks to my own safety.

Second-hand cars are exported from Japan as they have a limited local market there and yet can be re-sold across southern Africa (Dobler 2008). The international trade in used vehicles is a difficult topic to research because there are many attempts made by both exporters and importers to circumvent market controls such as taxes, customs charges, and vehicle regulations. Furthermore 'it is not always possible to quote sources of certain information' as trade activities can be 'very profitable, somewhat informal and therefore [information is] rather jealously guarded' (Nieuwenhuis *et al.* 2007: 18).

Since concealed, corrupt, and illegal activities were all involved in the import of used cars into Mozambique, I felt that it was important that some of the information I gleaned should not be revealed. Although the primary audience for my findings was academics, there was the potential for other traders or the authorities to access the information, and I was subsequently contacted by a car trader in Japan about my research findings. Moreover, specific details about transactions could directly damage individuals' livelihoods or even threaten people's safety (see also Fagerholm in this volume on the potential dangers of revealing local knowledge). The risk of the latter may seem distant but, as will be discussed below, acts of extreme violence have been linked to the used-vehicle trade in southern Africa.

Before beginning the main field research, I investigated the overall structure of the network which extends from Japan to Mozambique via South Africa using 'gray sources' (media stories and company reports). Following this, I undertook preliminary research by visiting car yards, viewing vehicles, and having brief interviews with various traders over several weeks. I used ethnographic methods to act as a 'participant' in the trade by posing as a customer. In such instances, I was performing a role which was expected of an expatriate living and working in Maputo; indeed during this time I did actually purchase a vehicle that I required to support my other work in Mozambique. This enabled me to glean a general understanding of the trade pattern through an inductive empirical approach, but also allowed me to build up contacts with car traders and spend time establishing trust.

Following this initial phase of research, I began 'hanging out' with traders and having casual discussions about the car trade; this became an increasingly important and deliberate research strategy and I recorded findings and observations in a fieldwork diary (Gardner 1999). I also interviewed used-car suppliers, importers, and customers.

Overall, my methodology followed Beuving's (2004) anthropological research of the used-car trade in Benin: through initial scoping, followed by repeated observation of core and standard processes, I established a detailed overview of the trade network. These observations included documenting processes that occurred in public spaces, such as at customs posts, where it would not have been appropriate to approach people for interview as it was suspected that bribes were being paid. Although there are known and well-documented difficulties with and limitations to the subjectivity of observational research (see Hoggart *et al.* 2002), comparable methods had previously been used in Mozambique by Sheldon (2003) and for this difficult topic this mode of research provided vital contextual information.

The subsequent focus of the fieldwork, which formed the main empirical material for an article that I wrote, was a series of transactions I followed in Mozambique and South Africa (Brooks 2012b). I accompanied a Mozambican used-car importer on a two-day business trip from Maputo to Durban and closely followed the processes of importing three different vehicles and the multiple transactions that he undertook during this journey. The trader willingly consented to the research process and I made a contribution to the travel expenses.

This in-depth ethnography gave me a detailed and invaluable insight into the corrupt practices that occur within the used-car trade, which would have been otherwise unobtainable. Implementing this method was only possible because I had been living in Mozambique and had the time and opportunity to build up a detailed understanding of the local second-hand car economy, which enabled me to progress from lay to expert knowledge and more importantly to establish trust and social contacts with a trader. When writing up the research, I took care in what I reported. I did not give details of individual interviews, trade deals, or the people involved. Anonymity for informants was very important and in this project it extended to not specifying particular fees, payments, or vehicles through which people could have been identified.

The risks associated with both participating in and researching the used-car trade were most pertinently demonstrated by the murder of Orlando José, a senior Mozambican customs official who had investigated the illegal import of cars during the same period when I was researching the second-hand car trade ('Ordem para matar terá vindo de dentro' 2010) (see also Tomei in this volume on research in dangerous places). Following this brutal killing, I carried out my research very cautiously and decided to limit the publication of any material through which participants could be identified. Understanding informal processes and gaining accurate insights into 'underground' economic activity is inherently difficult and my article was based on a small sample, precisely because gaining trust and exploring illegal trade activities is difficult (Harriss 1993; Thomas 1993) (see also Taylor in this volume on research among illegal squatters).

Despite the challenges of this project and the growing constraints on field research – increasingly enforced by ethical review procedures and risk assessment exercises (see Dyer and Demeritt 2008) – the illegalities and corruption in trade networks are realities and social problems with which critical social science research has to engage. This type of research is hard, and it is important to acknowledge, without compromising the quality of analysis, that there will be gaps in the information that is collected. Despite these limitations, I believe that research in countries of the Global South should continue to seek to provide insights in to the most pressing of social problems wherever possible, even if realistic, practical constraints limit what data can be gained.

Illegal activities: labor disputes and Chinese investment in Zambia

Chinese investment in Africa has attracted much recent critical work (Brautigam 2009). However, researchers have found it challenging to access this topic; the majority of publications have been on the international geopolitical scale and have involved macro-analysis of trends, rather than being based on in-depth field research (Carmody *et al.* 2011). The focus in discourse has been on the controversies relating to human rights, environmental impacts, access to natural resources, and – to a lesser extent – the treatment of labor. I chose to research the latter as I believed it was an important social issue that had been neglected. As Lee (2009: 4) notes, Chinese business in Africa has been criticized for being 'notorious in casualizing its workforce' and for paying the lowest wages.

I investigated Chinese investment through a case study which explored the tensions between labor and management in a clothing factory. In contrast to my research on second-hand car trading, which anonymized participants, in this study I named the specific factory – Zambia China Mulungushi Textiles (ZCMT) – and its parent company, Qingdao Textiles Corporation. I chose to do this because it would be impossible to 'anonymize' this case study, as the factory was a nationally important industrial site and easily recognizable to those familiar with the Zambian economy. Furthermore, and more fundamentally, my study dealt with the repercussions of the actions of a large corporate actor rather than individual people, and I believe that firms operating in southern Africa frequently escape critical attention and that such enterprises should bear greater accountability.

This chapter opened with quotes from two of my interviewees at ZCMT. The factory had been forced to close after the business became unviable. This had left many ex-workers unhappy following what they perceived to be their ill-treatment and the promises made and then broken by Chinese managers. The recent history of the factory had been marked by labor struggles which were influenced by race and gender (Brooks 2010). For my master's degree dissertation, I investigated the labor regimes at ZCMT through 21 interviews with Zambian ex-employees, which were facilitated by a research assistant.

The main challenge with this research was the emotional toll of dealing with impoverished and disillusioned people for whose plight there did not appear to be any ready solutions. Hearing stories of long hours, relentless work, difficult conditions, industrial accidents, and unemployment was deeply distressing (see also the chapters by Day and Skinner in this volume on the emotional impact of data collection from vulnerable people). I had substantially underestimated the likely impact upon myself of dealing with this challenge. I felt that the best approach for me was to try and work toward using the research participants' own words and phrases in my writing to communicate their story effectively (see also Gent in this volume on debates surrounding translation and representation). When I subsequently presented and wrote up this research, I was then able to use their voices to embellish the narrative and, after a time, move on from this difficult topic.

When preparing for this research project, I had been preoccupied by the practical details and methodological challenges, such as the need to acquire an adequate sampling frame. I had decided upon a snowball sampling technique, because I anticipated that finding and accessing the retrenched employees would be problematic: previous studies had illustrated that Zambian workers often 'retreated' from large settlements following the loss of employment (Ferguson 1999, Potts 2005).

However, in contrast to Ferguson's and Potts' experiences, I found that locating ex-workers to interview was relatively easy; many still lived in settlements adjacent to the factory as they hoped for compensation payments or opportunities for future re-employment on better terms. In fact, the main issue with this research related to the framing of the subject. The study (Brooks 2010) investigated Chinese engagement from the African perspective: it explored Zambians' experiences. When I designed the research in the UK – with limited knowledge and no experience of the issue – I had planned to conduct interviews with Chinese managers as well, to provide further perspectives. However, once I arrived, I found that the Chinese people who had been associated with the Mulungushi factory had left the area and I was unable to access other Chinese elsewhere in Zambia to provide another viewpoint.

In contrast, Lee (2009) was more successful in engaging both Africans and Chinese in fieldwork in Zambia as well as Tanzania. This was achieved through drawing on her own positionality and language ability as a Chinese-American, which helped her gain access to the enclaved Chinese populations. Equally in Ghana, Giese and Thiel (2012: 3) present a 'two-sided account of both the Ghanaian employees' and the Chinese employers' perspectives on the employment situation,' based on ethnographic research which was rooted in their dual regional specializations in West Africa and China.

These two studies offer wider perspectives than my own work, but I would argue that valid social research can still be undertaken from a single vantage point, especially when working within realistic constraints such as access to prospective participants. Research is not a neutral process and it is naive to think otherwise. Through discourse analysis we can untangle what we perceive to be the truths within our sample data, rather than merely presenting different sides to the same story. As Boykoff and Boykoff (2004) have expertly noted with respect to the media framing of climate change debates, 'balance is bias,' and presenting both sides of a politicized issue does not always make for the most accurate representation of reality. As researchers, we need to constantly think critically about the source material upon which we draw to make conclusions.

Making the case for intensive qualitative research

Development research has been criticized for following an extractive model of 'data mining' and having a profound inability to address the pressing demands and needs of research participants (Ferguson 1999; Hart 2002). This is especially true for research which involves controversial, corrupt, and illegal activities. Successful research projects frequently do nothing to address social problems – indeed, my own work has had very limited direct impact on people in Mozambique, South Africa, or Zambia – yet academics' achievements continue to be measured by publications and citations, whereas for postgraduates the criteria for success are final grades and completed dissertations.

In the field, we can consider measuring success in other ways. The dichotomy associated with cross-cultural research can be alleviated by the researcher offering to share stories and inviting questions about their own life to build a sense of rapport before and after interviews (see also Godbole in this volume on how much personal information to reveal or conceal). This can be personally rewarding and bring a different type of achievement from research interventions. Equally I am proud that I have tried to research topics which otherwise escape attention, and I have taken satisfaction in bringing difficult topics to an academic audience (Hall 2012).

Personal engagement may also facilitate research into difficult topics, as I found with my used-car research, but when undertaking prolonged fieldwork in the Global South it is important to maintain a critical distance, especially if informants are engaged with potentially unethical or illegal activities. When I undertook research in Maputo, the institutional support provided by the Instituto de Estudos Sociais e Económicos, which hosted my research, was invaluable as it provided a detached space for reflection and analysis away from 'the field.' Rabinow (1977/2007: 38) says that 'fieldwork is a dialectic between reflections and immediacy' and *apartness* in a different type of space allows us to recast and collect our thoughts and ideas.

In my experience, fieldwork does not present a succession of coherent encounters; rather each day brings different experiences and information which may or may not be useful. One good interview does not always lead to another. Moreover, I have found that the seemingly irrelevant data gained at the beginning of a research trip can actually prove to be useful several months later. Information is 'reconstructed' from research processes, and different episodes of fieldwork can be analyzed and linked together to provide a coherent narrative of events (Rabinow 1977/2007).

I believe that extended periods of fieldwork in different cultural contexts allow insights to be developed in the vein of ethnographic inquiry and that immersion in developing countries is vital to understanding the conditions of poverty (Iyenda 2007). The lived experience of being in southern Africa among market traders, car dealers, and ex-factory workers challenged my prior assumptions about 'other cultures' and enabled me to give more accurate perspectives on their socio-economic circumstances and the problems different people face (Hammersley 1992; Watson 2003). Being among people is important, but in effective research more has to be done than simply asking questions or collecting anecdotes; responses need to be assessed critically and positioned within a wider context (Hart 2002; Sender *et al.* 2006).

Academic preparedness is vital before embarking upon fieldwork, but in my experience it is as – if not more – important to be prepared for the personal as well as the methodological challenges. Risk assessments and ethical procedures can be useful, but in my opinion these constrain research into difficult topics and can deter early-career researchers from investigating pressing social problems. If before every research activity we were to weigh up *all* the potential consequences for ourselves and the research subjects, consider them earnestly and think about the immediate outcomes, then the probable repercussions, then the possible effects, and then finally the ultimately imaginable products of our actions, we would never set foot outside the library.

Practical constraints did of course affect what methods I applied in my research. I was an 'outsider' in most of my field settings. In some of those contexts I had agency which was helpful in managing the research activity (Skinner 2008). Foreign status can also make people reluctant to engage with you – or even hostile – as I experienced when attempting to interview Indian traders in Maputo (see also Hanlon 2009; see Dam and Lunn in this volume for the opposite 'outsider' experience). The examples of Lee (2009) and Giese and Thiel (2012) show how prior experience and cultural familiarity can facilitate research.

Research is also a form of power, and my positionality affected the outcomes. One of the reasons why market traders and ex-workers were comparatively easy to research in Mozambique and Zambia was because they were curious about the research and were accessible and interested in participating due to the 'surplus of spare time' which is common among the poor (Rabinow 1977/2007: 34). Underemployment is an unfortunate phenomenon, but can be an aid to the researcher. In contrast, when researching socio-economic problems it is important to consider that difficult-to-access participants may strive to conceal their activities and seek to avoid any potential publicity.

Through these different case studies I have attempted to explore the challenges of researching processes that are informal, illegal, unregulated, or concealed and which perpetuate inequality. These are mired in ethical dilemmas and practical difficulties, but I feel that research in this vein should be encouraged as it can be personally rewarding as well as successful in uncovering the causes and consequences of pressing social problems.

Acknowledgments

Thanks go to my MSc and PhD supervisors, Alex Loftus and David Simon, who provided invaluable advice on methodologies; and to the Instituto de Estudos Sociais e Económicos in Maputo, which hosted my research in Mozambique. I am also grateful for the support of research assistants, especially Manuel Ngovene, Helder Gudo, and Humphrey Zulu, who facilitated my fieldwork in southern Africa.

Recommended reading

Chambers, R. (1983) *Rural development: Putting the last first*, Harlow: Prentice Hall.

Chambers provides important lessons on the difficulties and potential biases of research in developing countries as well as the importance of focusing on the plight of the most impoverished.

Ferguson, J. (1999) *Expectations of modernity: Myths and meanings of urban life on the Zambian Copperbelt*, Berkeley, CA: University of California Press.
Ferguson, J. (2007) *Global shadows: Africa in the neoliberal world order*, Durham, NC: Duke University Press.

Ferguson's work in southern Africa provides excellent ethnographic studies, especially interesting for anyone wishing to learn more about the challenges of fieldwork in the region.

Hart, G. (2002) *Disabling globalization: Places of power in post-Apartheid South Africa*, Berkeley, CA: University of California Press.

This is an inspirational study into a difficult-to-research topic.

Rabinow, P. (1977/2007) *Reflections on fieldwork in Morocco: 30th anniversary edition*, Berkeley, CA: University of California Press.

An engaging and highly readable discussion of the messiness of cross-cultural research, Rabinow's *Reflections* remains a landmark.

References

Baden, S. and Barber, C. (2005) *The impact of the second-hand clothing trade on developing countries*, Oxford: Oxfam.
Beuving, J. J. (2004) 'Cotonou's Klondike: African traders and second-hand car markets in Bénin,' *Journal of Modern African Studies*, 42 (4): 511–537.
Boykoff, M. T. and Boykoff, J. M. (2004) 'Balance as bias: Global warming and the US prestige press,' *Global Environmental Change*, 14 (2): 125–136.

Brautigam, D. (2009) *The dragon's gift: The real story of China in Africa*, Oxford: Oxford University Press.

British Heart Foundation (2011) 'Charitable public misled by commercial collection companies', 31 August 2011, available at: www.bhf.org.uk/media/news-from-the-bhf/charitable-public-misled.aspx (accessed: 5 September 2011).

Brooks, A. (2010) 'Spinning and weaving discontent: Labour relations and the production of meaning at Zambia–China Mulungushi Textiles,' *Journal of Southern African Studies*, 36 (1): 113–132.

Brooks, A. (2012a) 'Riches from rags or persistent poverty? The working lives of second-hand clothing vendors in Maputo, Mozambique,' *Textile: The Journal of Cloth and Culture*, 10 (2): 222–237.

Brooks, A. (2012b) 'Networks of power and corruption: The trade of Japanese used cars to Mozambique,' *The Geographical Journal*, 178 (1): 80–92.

Brooks, A. (2012c) *Riches from rags or persistent poverty? Inequality in the transnational second-hand clothing trade in Mozambique*, unpublished thesis, Royal Holloway University of London.

Brooks, A. (2013) 'Stretching global production networks: The international second-hand clothing trade,' *Geoforum*, 44: 10–22.

Brooks, A. and Simon, D. (2012) 'Untangling the relationship between used clothing imports and the decline of African clothing industries,' *Development and Change*, 43 (6): 1265–1290.

Carmody, P., Hampwaye, G. and Sakala, E. (2011) 'Globalisation and the rise of the state? Chinese geogovernance in Zambia,' *New Political Economy*, 17 (2): 209–229.

Chataway, J., Joffe, A. and Mordaunt, J. (2007) 'Communicating results,' in A. Thomas and G. Mohan (eds), *Research skills for policy and development: How to find out fast*, London: Sage.

Davis, M. (2004) 'Planet of slums: Urban involution and the informal proletariat,' *New Left Review*, 1 (26): 5–34.

Dobler, G. (2008) 'From Scotch whisky to Chinese sneakers: International commodity flows and new trade networks in Oshikango, Namibia,' *Africa: The Journal of the International African Institute*, 78 (3): 410–432.

Dyer, S. and Demeritt, D. (2008) 'Un-ethical review? Why it is wrong to apply the medical model of research governance to human geography,' *Progress in Human Geography*, 33 (1): 64–64.

Ferguson, J. (1999) *Expectations of modernity: Myths and meanings of urban life on the Zambian Copperbelt*, Berkeley, CA: University of California Press.

Fife, W. (2005) *Doing fieldwork: Ethnographic methods for research in developing countries and beyond*, New York: Palgrave Macmillan.

Gardner, K. (1999) 'Location and relocation: Home, "the field" and anthropological ethics (Sylhet, Bangladesh),' in C. W. Watson (ed.) *Being there: Fieldwork in anthropology*, London: Pluto Press.

Giese, K. and Thiel, A. (2012) 'The vulnerable other: Distorted equity in Chinese–Ghanaian employment relations,' *Ethnic and Racial Studies*, DOI: 10.1080/01419870.2012.681676.

Gladwin, C. H., Peterson, J. S. and Mwale, A. C. (2002) 'The quality of science in participatory research: A case study from Eastern Zambia,' *World Development*, 30 (4): 523–543.

Hall, T. (2012) 'Geographies of the illicit: Globalization and organized crime,' *Progress in Human Geography*, DOI: 10.1177/0309132512460906.

Hammersley, M. (1992) *What's wrong with ethnography?* London: Routledge.

Hanlon, J. (2009) 'Mozambique: The panic and rage of the poor,' *Review of African Political Economy*, 36 (119): 125–130.

Hansen, K. T. (2000) *Salaula: The world of secondhand clothing and Zambia*, Chicago: Chicago University Press.

Harriss, B. (1993) 'Talking to traders about trade,' in S. Devereux and J. Hoddinott (eds), *Fieldwork in developing countries*, Boulder, CO: Lynne Rienner Publishers.

Hart, G. (2002) *Disabling globalization: Places of power in post-Apartheid South Africa*, Berkeley, CA: University of California Press.

Hoggart, K., Lees, L. and Davies, A. (2002) *Researching human geography*, London: Arnold.

Iyenda, G. (2007) 'Researching urban poverty in sub-Saharan Africa,' *Development in Practice*, 17 (1): 27–38.

Kamete, A. Y. and Lindell, I. (2010) 'The politics of "non-planning" interventions in African cities: Unravelling the international and local dimensions in Harare and Maputo,' *Journal of Southern African Studies*, 36 (4): 889–912.

Kapferer, B. (1972) *Strategy and transaction in an African factory*, Manchester: Manchester University Press.

Kitchin, R. and Fuller, D. (2003) 'Making the "black box" transparent: Publishing and presenting geographic knowledge,' *Area*, 35 (3): 313–315.

Kleine, D. (2008) 'Negotiating partnerships, understanding power: Doing action research on Chilean fairtrade wine value chains,' *Geographical Journal*, 174 (2): 109–123.

Latour, B. (1993) *We have never been modern*, Cambridge, MA: Harvard University Press.

Lee, C. K. (2009) 'Raw encounters: Chinese managers, African workers and the politics of casualization in Africa's Chinese enclaves,' *The China Quarterly*, 199: 647–666.

Le Mare, A. (2007) 'Case study: Fair Trade and producer livelihoods – Exploring the connections', Economic and Social Research Council, available at: www.esrc.ac.uk/_images/Case_study_LeMare_tcm8–2340.pdf (accessed 27 July 2011).

Manzo, L. C. and Brightbill, N. (2007) 'Toward a participatory ethics,' in S. Kindon, R. Pain and M. Kesby (eds), *Participatory action research approaches and methods: Connecting people, participation and place*, London: Routledge.

Nieuwenhuis, P., Beresford, A. and Young Choi, K. (2007) 'Shipping air? Tracking and forecasting the global shipments of new and used cars,' *Journal of Maritime Research*, 4 (3): 17–36.

'Ordem para matar terá vindo de dentro' (2010) *Savana*, 30 April.

Paasche, T. F. and Sidaway, J. D. (2010) 'Transecting security and space in Maputo,' *Environment and Planning A*, 42 (7); 1555–1576.

Pitcher, A. (2002) *Transforming Mozambique: The politics of privatization, 1975–2000*, Cambridge: Cambridge University Press.

Potts, D. (2005) 'Counter-urbanization on the Zambian Copperbelt? Interpretations and implications,' *Urban Studies*, 42 (4): 583–609.

Rabinow, P. (1977/2007) *Reflections on fieldwork in Morocco: 30th anniversary edition*, Berkeley, CA: University of California Press.

Sender, J., Oya, C. and Cramer, C. (2006) **'Women working for wages: Putting flesh on the bones of a rural labour market survey in Mozambique,'** *Journal of Southern African Studies*, 32 (2): 313–333.

Sheldon, K. (2003) 'Markets and gardens: Placing women in the history of urban Mozambique,' *Canadian Journal of African Studies*, 37 (2/3): 358–395.

Skinner, C. (2008) *Street trade in Africa: A review*, School of Development Studies, Working Paper No. 51, Durban: University of KwaZulu-Natal.

Smith, M. (2004) 'Contradiction and change? NGOs, schools and the public faces of development,' *Journal of International Development*, 16 (5): 741–749.

Thomas, J. J. (1993) 'Measuring the underground economy: A suitable case for interdisciplinary treatment?,' in C. M. Renzetti and R. M. Lee (eds), *Researching sensitive topics*, London: Sage.

Watson, E. E. (2003) ' "What a dolt one is": Language learning and fieldwork in geography,' *Area*, 36 (1): 59–68.

Zarkovich, S. S. (1993) 'Some problems of sampling work in underdeveloped countries,' in M. Bulmer and D. P. Warwick (eds), *Social research in developing countries*, London: Routledge.

5 Finding fluency in the field: ethical challenges of conducting research in another language

Experiences from fieldwork on the solar energy sector in Nicaragua

Danielle Gent

> Learning how to speak Nica Spanish, which differs greatly from Spanish Spanish, and again from Honduran Spanish has been a challenge. These first few weeks in the field have been unbelievably frustrating! I understand almost everything, but at times feel completely unable to express myself. I think that this has affected how people perceive me ... I arrived at the NGO and became flustered when I tried to explain my research objectives. I think this gave the impression of an inexperienced and underprepared researcher, who needed guidance.

My ability to speak Spanish – and to speak it well – was one important factor in my choice of fieldwork location. The capacity to speak the local language, it has been argued, is the 'centrepiece of successful field research' (Veeck 2001: 34). Similarly, even if the researcher's language proficiency is not perfect, participants will respect researchers who make the effort to 'penetrate their way of speaking' (Gade 2001: 377). Although the literature makes it sound relatively straightforward, in practice, conducting research in a second language can be difficult and frustrating. It can, however, also be incredibly rewarding and inevitably enriches our field experiences.

Closely related to language is the issue of representation, and while all researchers grapple with the issue of how to represent voices in qualitative research, this becomes decidedly hazier when research is conducted in a second language. As Temple (2005) suggests, there is no single correct way for researchers to represent people who speak in different languages. Yet this raises the question of how we reconcile the tension arising from the fact that often the operational (second) language of fieldwork differs from the (home) language of final publication or monograph. The choices we make about representation have political, epistemological, and ethical implications. These issues aside, overseas fieldwork characteristically demands local language proficiency or the use of an interpreter, meaning that the researcher may face a plethora of additional practical and ethical considerations, difficulties, and anxieties.

This chapter reflects on some of the linguistic challenges I encountered during field research in Nicaragua and the associated ethical issues. It focuses on three interrelated challenges of field research in a second language: the process of learning and becoming proficient in the field; moving and communicating between the different 'worlds' of research; and issues of translation and representation once the fieldwork phase is over.

Re-learning the language(s)

My research focused on the role of solar energy in Nicaragua, in particular the governance and sustainability of the sector and its uses at the local level. At the national level, I conducted semi-structured interviews with the 'elites' responsible for the design of energy policy and the implementation of the solar energy programs themselves. These were mostly people based in the capital city, Managua, who were well versed in the language of sustainable energy. At the local level, I conducted case study research across three solar home system programs. This involved spending time with individuals in remote rural communities, talking to them about their daily lived experiences of energy and what changes, if any, the arrival of solar energy had wrought in their lives.

Proficiency with the Spanish language was clearly critical, not only in enabling me to meet and address the research aims, but also for communicating with and understanding the perspectives of my research participants. Prior to embarking on my doctoral research, I had spent a year studying in Spain and doing voluntary work in Honduras. I therefore felt confident in my ability to speak Spanish, as well as being eager to return to a Spanish-speaking country. Indeed, I felt that choosing a field location where I had not only a level of fluency in the language but also an understanding of the culture would be beneficial for my research and would reduce the likelihood of 'culture shock.' One of my supervisors had contacts in Nicaragua, which had undergone some interesting developments in solar energy, and we decided that the country would provide a good case study.

On arrival in the field, however, I found that Nicaraguan Spanish was rather different from the Spanish I had learnt previously. I had not anticipated that communication difficulties would be the greatest cause of anxiety during the initial weeks and months of fieldwork. Indeed, I was almost required to re-learn my Spanish in order to understand the accents and vernaculars of different Nicaraguan actors and to feel confident that I understood the different voices that I was hearing and that I would be able to portray them with some accuracy.

With hindsight, it appears that it might have been beneficial to have – at this stage – undertaken formal language lessons; however, I was aware of my time in the field ticking away, and as my research began to pick up pace this ceased to be an option. Another alternative might have been to employ an interpreter, but I did not want to do this for several reasons. First, although I was having difficulties, I did speak Spanish! Second, as far as possible, I wanted to be a participant rather than a passive observer in the different worlds I was experiencing. Third, I believed that talking to participants directly, taking on their words and understanding the vocabulary of their everyday lives, was essential for my understanding. Indeed, over time, Nicaraguan Spanish became not only the way in which I conceptualized and critiqued my research, but also how I interpreted and understood the context within which research participants and myself were embedded.

While the language difficulties lessened as my field research progressed, they never truly disappeared. In my case, this was complicated by the fact that as I moved from my Nicaraguan home to the offices and villages of my participants, my everyday operating language shifted from my native English (and the largely Anglo-American literature of my research area, for example see Desbiens 2002; Rodríguez-Pose 2006; Hassink 2007) to Nicaraguan Spanish.

In addition to moving between languages and familiarizing myself with manifold accents, vernaculars, and vocabularies, conducting research in my second language also required learning to speak the language of energy, which also differed. While elite actors were conversant in a more technical language, those in rural areas, who had lived without electricity for the majority of their lives, spoke about energy less formally. The following quote from my field diary discusses the ways in which local people referred to electricity:

> I find it both fascinating and confusing that people here refer to their electricity bills as 'gastos de luz' or lighting bills. Rather than talking about the energy carrier itself, they define it as the service – lighting; grid electricity similarly is called 'luz convencional' or conventional lighting – again emphasizing the service, rather than the carrier.... Some interviews have been really confusing.

This example highlights how understanding the etymology of this use of the word *luz* was critical for me to understand how local people talked about and understood energy. I understood *luz* to mean 'light' in *castellano* Spanish. On asking a contact at the university why electricity was referred to as 'light' he expressed surprise, stating that he had not questioned it before, although he was familiar with its history. He explained that the Nicaraguan state energy company had been called Empresa Nacional de Luz y Fuerza (National Light and Power Company), and *luz* provided an informal, everyday way of talking about electricity. Perhaps if my research participants' words had been relayed to me by an interpreter, or had I been less proficient in Spanish, I would have missed out on this particular representation of electricity (see also Leck in this volume for a similar experience on local words for, and interpretations of, 'climate change').

As my time in the field progressed, I became more knowledgeable of the different vocabularies used by participants to discuss not just energy-related issues, but also their daily lives. In the next section I discuss the different 'worlds' I encountered during fieldwork and the ways in which I came to terms with moving within and between them, where the issue of language added a further layer of complexity.

Learning to communicate within and between research 'worlds'

> All the while I feel I'm living a double life – on week days I get to grips
> with rural people living in extremely difficult and impoverished conditions,
> interspersed by meetings with key decision makers in air conditioned offices
> in Managua, and when I get a day off, I take advantage of being in a beauti-
> ful country half way across the world … there really are many different
> 'worlds' to Nicaragua.

This excerpt from my field diary describes my feelings about living in and
between the different worlds in which I was carrying out my field research. This
challenge – experiencing different realities geographically close yet socially disparate – relates to the previous section in that it required familiarity with different dialects, accents, and vernaculars. With this also came an ethical dilemma in
my ability to (re)present myself and my research within and across the different
'worlds.'

In this respect, my greatest challenge was to understand participants in rural
communities, as well as to make myself and my research understood by these
participants. I returned from my first interviews in rural communities feeling
utterly downhearted, disappointed, and anxious. Despite following the advice
provided by Gade (2001) and Veeck (2001) – who suggest that the researcher
should arrive with a structured guide of almost 'scripted' questions in the source
language – and having what I thought was an interview schedule of grammati-
cally correct, comprehensible, and logically ordered questions, my phrasing and
choice of words, as well as the level of formality that I had adopted, turned out
to be incomprehensible to some participants. Many early interviews involved
participants looking bewildered at my accompanying NGO worker, who would
then 'translate' my question into more appropriate words. Upon sharing my concerns with my host, I was advised to 'denigrate' my Spanish.

While I heeded her advice, I also decided that having the support of a field
assistant would be beneficial, not just to aid communication, but also to address
logistical and safety issues. Jimmy, a social anthropologist and my field assistant, came to be both a friend and a cultural and linguistic 'broker' at interviews.
As a result of Jimmy's support, my interview experiences were greatly enriched.
As Bujra (2006) suggests, the presence of a research assistant – even if they
cannot translate the local vernacular into English – is valuable as they may be
able to offer an 'extended gloss' which facilitates a greater grasp of meaning. I
certainly found this to be the case. I also conducted a debrief at the end of each
interview, giving the interviews what I call 'two fingerprints'; this aided triangulation in that we could challenge each other's understandings and interpretations
of the interview (see also the chapters by Le Masson, Leck, and Lunn and
Moscuzza in this volume for similar use of triangulation with research assistants). Heller *et al.* (2011) agree, finding the debrief technique a useful experience for gaining another's perspective on issues raised.

Language, however, may not be the only separation between the different 'worlds' encountered during field research. Gade (2001), for example, discusses how a diversity of approaches was required across his different research contexts. He cites the example of research among predominately non-literate people, where participants did not grasp the sentiment of his interrogations; for example, in the Andes anyone wearing boots and carrying a map was referred to as *ingeniero* (engineer). Conversely in Europe, Gade found that the concept of research was one understood and accepted by participants and thus posed fewer difficulties to his studies. I encountered similar difficulties in presenting myself and my research to remote communities that were without previous experience of foreign researchers. I, the *gringa* (Yankee) or *chela* (fair skinned) *investigadora* (researcher), was at times perceived as a *cooperante* (aid worker). This could have profound implications on the data I produced, but it also led me to question my presence in such remote and relatively marginalized communities as a researcher. Why was I conducting research instead of practically engaging with communities to face their perceived issues? (See also the chapters by Chopra and Jones in this volume regarding the tensions between being both an academic and an activist or practitioner.)

Like Gade, I found that in the other 'world' the concept of research was well understood. Furthermore, language did not present the same challenge when I was conversing with the so-called elites of the Nicaraguan energy sector. Although these conversations presented other challenges related to access and power relations (e.g., Cormode and Hughes 1999), I felt more comfortable linguistically as I was able to converse with confidence in a 'common language' – that of sustainable energy. The Spanish spoken in the offices of Managua also had more in common with the more formal language I had learnt in Spain. With the elites, I also shared certain aspects of biography; these included racial, class, and educational backgrounds, which in most cases facilitated personal rapport with interviewees (Mais 2009; Oglesby 2010; see also the chapters by Chopra, Dam and Lunn, Perera-Mubarak, and Wang in this volume on interviewing elites). This relates to Herod's (1999) discussion of interviewing foreign elites, in which he argues that researcher 'insider'/'outsider' status (e.g., Mullings 1999) is far from a dichotomy but is more complex and fluid.

This section has demonstrated that moving within and between the different research worlds gives rise to ethical challenges relating to the ease of (re) presentation to our participants. Indeed, language may not be the only 'difference' between us, as Crane *et al.* (2009) found; in many instances the social and economic differences between researcher and informants are far more significant to field activities than linguistic barriers. This brings us to the final challenge I experienced during field research in my second language, that of translation, interpretation, and representation.

Ethical dimensions of representing voices and translating the field

> Outside the context of the fieldwork I had the power to determine not only how participants were represented to the outside world, but if they were represented at all. In effect, in common with all forms that utilize ethnography; the predominant mode of modern fieldwork authority is signalled; 'you are there ... because I was there.'
>
> (Linkogle 2000: 144)

The above quote from Linkogle nicely captures the final ethical dimension of discussion in this chapter – that is, the power of the researcher to represent 'others.' Indeed, as Twyman *et al.* (1999) suggest, these kinds of issues are further compounded once foreign language and translation are involved. The dilemma of representation applies not only to those researchers working in a second language, however, but to all qualitative researchers who through their dissemination must choose whose voices are represented, as well as how and to whom (for example see Bennett and Shurmer-Smith 2001; Crang 2001; Ley and Mountz 2001). Wasserfall (1993, cited in England 1994) argues that while researchers cannot fully represent the voices of their informants, they must take responsibility for this representation. England (1994) goes further, calling for an awareness of how the researcher's biography filters the information received, as well as his or her perceptions and interpretations of the fieldwork experience.

In my research these power relations were not always clear-cut. For example, when interviewing elites and non-elites, I was sometimes in a position of relative power and at other times this was reversed (see Wang in this volume for a similar experience in interviews with business people). However, as Linkogle's quote suggests, the researcher's power definitively comes to the fore when writing and representing. Indeed, in writing my monograph, I had the power to play the part of 'mediator' between the two languages of my research (Smith 1996), but also to 'transport people's spoken words, which are lively and used in everyday lives, into academic text' (Kim 2012: 138). This can be thought of as a 'dual transformation' – one not only of translation (into my native English), but also one of authorization (into academic prose), where I then faced the ethical dilemma of choosing how to deliver others' words to perceived audiences (Kim 2012).

On returning from fieldwork, my operational or everyday language of research shifted back to my native English, yet I found myself confronted with hundreds of pages of transcripts in Spanish. How was I to make sense of these without the frequent contact I had with participants and native speakers? I found relatively limited guidance on how I should go about data analysis in a second language, the results of which would then be presented in my native language. Having already attempted to translate transcripts into English and then to analyze them, I found that much of the richness of the original interview exchange was removed. As a consequence, I decided to follow the advice of Smith (2003) who suggests conducting interview analysis in the source language and only then translating excerpts. I found that this approach not only allowed me to maintain the richness and honesty of the original exchange, but also facilitated my connection with participants and the local context. Indeed, Twyman *et al.* (1999) argue that data should be analyzed with regard to the cultural context in which it was constructed, a process they call 'transculturation.'

Following Smith's advice, however, does not remove the difficulties of having to translate elements of the original transcript, which, according to Temple (2005), is a process fraught with political and ethical difficulties. Translation is often presented as a *fait accompli* (Frenk 1995, Temple 2005, Müller 2007) which conceals the difficulties of the translation process and assumes its neutrality. Indeed, the translating researcher faces enormous difficulties, due to the 'inadequation of one tongue to another' (Derrida 1991: 244). Furthermore, and as I experienced, 'any translation always seems to be a reduced and distorted representation of other social texts and practices' (Smith 1996: 162).

So how can researchers navigate the translation minefield, bearing in mind that there is no correct or accepted way to do it? While recognizing the impossibility of revealing the 'truth' of the 'other' in the 'home' language (e.g., Smith 1999), various authors provide advice as to how this can be negotiated that I found to be particularly helpful in my own writing. Hassink (2007: 1286), for example, pleads for 'fewer translations: don't translate the untranslatable!' suggesting that authors should use glossaries to paraphrase key terms in the native language. Meyer and Maldonado-Alvarado (2010: 32) provide an excellent example of this; they found it difficult to adequately translate terms such as *comunalidad* and *interculturalidad* in their edited book, and so 'consistently leave these terms in their Spanish to mark … words and cultural concepts as untranslatable,' instead allowing different contributors to elaborate on their conceptualization of the terms. I also chose to adopt this technique, known as 'holus bolus' (Müller 2007), which enabled me to maintain the range of peculiar, particular, and colloquial expressions that were specific to my Nicaraguan participants. By maintaining source language expressions as 'markers of difference' when I came to write (ibid.: 210), I was able to keep the richness of the original text while drawing attention to the contingency of meaning; the example of *luz*, given earlier, is illustrative of this richness.

Second language research: well worth the challenge

Reflecting on my own experiences of research in a second language, I can honestly say that – despite the anxieties and frustrations – it was ultimately a rewarding and enriching experience. I found it immensely satisfying to become more proficient in Spanish (of various types!). In particular, the ability to communicate confidently with others from different countries and backgrounds, in their language, using their words and expressions, allowed me to experience worlds that I could not have done without this ability. However, the demands of doing research in a second language should not be underestimated; field research in the

Global South can be a testing and emotional experience and undertaking research in a second language inevitably brings with it additional challenges.

The ethical issue of representation – whether this is representing the research and researcher to participants within different 'worlds,' or the 'voices' of those participants once back at home – is a recurrent theme throughout this chapter and within the wider literature. How the researcher chooses to represent participants is ultimately a political and personal process, regardless of whether the research is conducted in a second language. Self-reflexive, context-conscious fieldwork allows the researcher to act transparently when grappling with such issues.

This chapter should be of relevance to other researchers in three ways: first, in showing interpretation and representation as inherently political and personal processes, requiring reflexivity and sensitivity by the researcher (England 1994; Punch 2012); second, in noting that undertaking research in a second language makes processes of interpretation and representation more complex; and third, in suggesting that transparency by the researcher can serve to minimize misrepresentation, as England (1994: 86) argues, making the researcher 'more aware of asymmetrical or exploitative relationships.'

Recommended reading

There are a great many references on the perils of field research in a second language. While many of the references cited in this chapter will be useful, I found the following particularly insightful during my own research:

Barley, N. (1983) *The innocent anthropologist: Notes from a mud hut*, London: Penguin.
Gade, D. (2001) 'The languages of foreign fieldwork,' *Geographical Review*, 91 (1–2): 370–379.
Veeck, G. (2001) 'Talk is cheap: Cultural and linguistic fluency during field research', *Geographical Review*, 91 (1–2): 34–40.
Watson, E. (2004) '"What a dolt one is": Language learning and fieldwork in geography,' *Area*, 36 (1): 59–68.

Gade (2001), Veeck (2001) and Watson (2004) all provide honest accounts of learning to speak and operate in the language of the field; Barley (1983) provides a humorous account of anthropological fieldwork in Cameroon, where one of the challenges he faced was to learn the local dialect.

Professor Bogusia Temple has written extensively on the issues of language and translation in sociological research; a list of her publications is available at: www.uclan.ac.uk/schools/school_of_social_work/research/bogusia_temple_tab_profile.php.

Punch, S. (2001) 'Household division of labour: Generation, gender, age, birth order and sibling composition,' *Work, Employment & Society*, 15 (4): 803–823.

Dr Samantha Punch (2001) provides an alternative strategy to the 'holus bolus' technique, which involves providing the original and translated quotation. This offers the reader the opportunity to critically assess the author's translation against their own interpretations. Her writings also provide honest and sincere accounts of the often hidden struggles of fieldwork. A list of her publications is available at: www.dass.stir.ac.uk/staff/Dr-Samantha-Punch/51.

References

Bennett, K. and Shurmer-Smith, P. (2001) 'Writing conversation,' in C. Dwyer and S. Limb (eds), *Qualitative methodologies for geographers: Issues and debates*, London: Arnold.

Bujra, J. (2006) 'Lost in translation? The use of interpreters in fieldwork,' in V. Desai and R. Potter (eds), *Doing development research*, London: Sage.

Cormode, L. and Hughes, A. (1999) 'Editorial introduction: The economic geographer as a situated researcher of elites,' *Geoforum* 30 (4): 299–300.

Crane, L. G., Lombard, M. B. and Tenz, E. M. (2009) 'More than just translation: Challenges and opportunities in intercultural and multilingual research,' *Social Geography Discussions*, 5: 51–70.

Crang, M. (2001) 'Field work: Making sense of interview data,' in C. Dwyer and S. Limb (eds), *Qualitative Methodologies for Geographers: Issues and Debates*, London: Arnold.

Derrida, J. (1991) *A Derrida reader: Between the blinds*, Hemel Hempstead: Harvester Wheatsheaf.

Desbiens, C. (2002) 'Speaking in tongues, making geographies', *Environment and Planning D: Society and Space*, 20 (1): 1–25.

England, K. V. L. (1994) 'Getting personal: Reflexivity, positionality and feminist research,' *The Professional Geographer*, 46 (1): 80–89.

Frenk, S. (1995) 'Re-presenting voices,' in J. G. Townsend, *Women's voices from the rainforest*, New York: Routledge.

Gade, D. (2001) 'The languages of foreign fieldwork,' *Geographical Review*, 91 (1–2): 370–379.

Hassink, R. (2007) 'It's the language stupid! On emotions, strategies, and consequences related to the use of one language to describe and explain a diverse world', *Environment and Planning A*, 39: 1282–1287.

Heller, E., Christensen, J., Long, L., Mackenzie, C. A., Osano, P. M., Ricker, B., Kagan, E. and Turner, S. (2011) 'Dear Diary: Early career geographers collectively reflect on their qualitative field research experiences,' *Journal of Geography in Higher Education*, 35 (1): 67–83.

Herod, A. (1999) 'Reflections on interviewing foreign elites: Praxis, positionality, validity and the cult of the insider,' *Geoforum*, 30: 313–327.

Kim, Y. J. (2012) 'Ethnographer location and the politics of translation: Researching one's own group in a host country,' *Qualitative Research* 12: 131–146.

Ley, D. and Mountz, A. (2001) 'Interpretation, representation, positionality: Issues in field research in human geography,' in C. Dwyer and S. Limb (eds), *Qualitative methodologies for geographers: Issues and debates*, London: Arnold.

Linkogle, S. (2000) '*Relajo*: danger in the crowd,' in G. Lee-Treweek and S. Linkogle (eds), *Danger in the field: Risk and ethics in social research*, London: Routledge.

Meyer, L. and Maldonado-Alvarado, B. (eds) (2010) *New world of indigenous resistance: Noam Chomsky and voices from North, South, and Central America*, San Francisco: City Lights.

Mais, T. R. D. (2009) *Transforming development? The Millennium Challenge Account and US–Nicaraguan relations*, unpublished thesis, Loughborough University.

Müller, M. (2007) 'What's in a word? Problematizing translation between languages,' *Area*, 39 (2): 206–213.

Mullings, B. (1999) 'Insider or outsider, both or neither: Some dilemmas of interviewing in cross-cultural settings,' *Geoforum*, 30: 337–350.

Oglesby, E. (2010) 'Interviewing landed elites in post-war Guatemala,' *Geoforum*, 41: 23–25.

Punch, S. (2001) 'Household division of labour: Generation, gender, age, birth order and sibling composition,' *Work, Employment & Society*, 15 (4): 803–823.

Punch, S. (2012) 'Hidden struggles of fieldwork: Exploring the role and use of field diaries,' *Emotion, Space and Society*, 5: 86–93.

Rodríguez-Pose, A. (2006) 'Is there an "Anglo-American" domination in human geography? And, is it bad?,' *Environment and Planning A*, 38 (4): 603–610.

Smith, F. (1996) 'Problematising language: Limitations and possibilities in "foreign language" research,' *Area*, 28 (2): 160–166.

Smith, F. (1999) 'The neighbourhood as site for contesting German reunification,' in J. Sharp, P. Routledge, C. Philo and R. Paddison (eds), *Entanglements of power: Geographies of dominance and resistance*, London: Routledge.

Smith, F. (2003) 'Working in different cultures,' in N. Clifford and G. Valentine (eds), *Key methods in geography*, London: Sage.

Temple, B. (2005) 'Nice and tidy: Translation and representation,' *Sociological Research Online*, 10 (2).

Twyman, C., Morrison, J. and Sporton, D. (1999) 'The final fifth: Autobiography, reflexivity and interpretation in cross-cultural research,' *Area*, 31 (4): 313–325.

Veeck, G. (2001) 'Talk is cheap: Cultural and linguistic fluency during field research,' *Geographical Review*, 91 (1–2): 34–40.

6   Whose voice? Ethics and dynamics of working with interpreters and research assistants

Experiences from fieldwork on climate change adaptation in South Africa

Hayley Leck

> This morning Sibongile and I met with the Nkosi[1] at Ngcolosi for permission to do research in his community. She spoke to him with such ease and confidence (despite telling me how nervous she was). Thank goodness for Sibs, I would be lost without her help in situations like today!

My fieldwork accomplice, Sibongile Buthelezi, acted as both my interpreter and my research assistant. The above excerpt from my research diary highlights one of many instances when Sibongile's assistance proved critical to the success of my research. Not only did she interpret and translate interviews, surveys, and conversations, but her social skills, local knowledge, and insights proved invaluable to my research, as did her emotional support.

While there are certainly differences between a research assistant (RA) and an interpreter or translator, and between the degrees of involvement that each implies, I apply the terms interchangeably in this chapter because in my case Sibongile played both these roles (and many more). Sibongile also viewed her role in this light. Therefore, in accordance with both our preferences, I introduced her to people as my research assistant and referred to 'our' research during interactions (see also Staddon in this volume for a similar relationship with her RA).

Participants thus viewed Sibongile as my assistant, with an active role in the research. Further, when Sibongile was doing research in my absence, it is likely that some participants viewed her as an equal actor in the research as she was the key point of contact in such instances. It is important to discuss this issue of identity construction at the outset of fieldwork in order to clarify respective understandings of roles and responsibilities in the research process and the implications of these.

Despite their significance, research assistants' personal perspectives and experiences are rarely documented when researchers consider the ethical dimensions of their work. In other words, these voices are often largely absent from oral and written accounts, including in dissertations and theses. In this chapter I reflect critically on the ethical aspects of doing research with interpreters or research assistants from a distinctive standpoint that underscores the centrality of accounting for their critical 'voice(s)' at various stages of the research process.

In the following section I set out the range of ethical challenges involved with working with interpreters and RAs. Following this, I reflect critically on my personal experience of working with an RA and some of the ethical dilemmas that we faced. I attempt to include Sibongile's voice in these reflections to underscore her key contributions throughout the research process. The chapter concludes with a few suggestions for dealing with some of the ethical dimensions of working with RAs or interpreters and facilitating the effective representation of their voices and experiences.

A multitude of ethical considerations

Many researchers who undertake research in 'foreign' contexts rely, to varying degrees, on the support of RAs or interpreters for doing fieldwork. They not only help to overcome the language barrier but can also provide insight on local community dynamics and guidance on cultural and safety issues, and may negotiate access to local gatekeepers. Despite this, and notwithstanding the widespread importance ascribed to reflexivity and context considerations in qualitative research, RAs and interpreters are often peculiarly omitted from oral and written research accounts (Temple and Edwards 2002).

I would like to argue that more attention needs to be given to the ethical responsibility for principal researchers to consider critically their RAs' views on the research and to make them more visible in the research process and outcomes. After all, in many cases, RAs and interpreters have the most direct interaction with research participants and therefore have critical insights into research findings and stand to make key contributions to research.

My doctoral research focused on responses to environmental change, or more narrowly climate change, at municipal and household levels in two neighboring local government districts in KwaZulu-Natal province, South Africa. As I was interested in people's understandings of, perceptions of and responses to climate variability and change, I needed to hear and understand the 'voices' of research participants (Hennink *et al.* 2011). However, although I was doing research at 'home' in my native South Africa, I did not have enough capability in the local language of all the communities that I was researching; hence the need to engage Zulu-speaking RAs for four of my ten case study communities to assist with the interviews, surveys, and focus groups.

It was shortly after starting fieldwork that I realized that I had given inadequate ethical consideration to my RAs. In preparation for fieldwork I had considered some ethical components such as in-the-field safety concerns for my RAs, but I had not anticipated, nor contemplated adequately, the potential impacts (positive or negative) of my study on my RAs or the importance of accounting for their voice throughout the research process.

RAs have personal opinions regarding their role and influence on the research process that can challenge and extend our own (Meth 2003). RAs' voices are also inflected with wider sets of power relations; RAs bring their own concerns, values, beliefs, and assumptions to the research process which are shaped by their own social worlds and require consideration. Indeed, RAs bring important and unique insights to our research and results, and as Temple and Edwards (2002) cogently explain, when working with RAs, research is influenced by 'triple subjectivity' (the researcher/research assistant/research participant relationship), which requires explicit consideration.

As detailed in the remainder of this chapter, I have grappled with various complex and multifaceted ethical aspects of working with RAs, some of which I feel I have dealt with appropriately and others less so. In the following sections, I attempt to incorporate Sibongile's voice into my own critical reflections of some key research aspects and ethical dilemmas that we faced and to highlight the key contributions that she made through all stages of the research process.

Choosing the right person

The selection and training of research assistants and interpreters is crucial: 'A good assistant can make or break one's research and one's relations with the local community, even for local researchers' (Apentiik and Parpart 2006: 34). I appointed Sibongile as my main RA for working with Zulu-speaking communities. I first met her in the third year of my undergraduate degree, when she was assisting on various research projects within the department. Sibongile's extensive research experience and high recommendations from local colleagues were key deciding factors when I was seeking a RA.

Sibongile is a black, educated South African woman and does research work on a contract basis within the University of KwaZulu Natal. She is a friendly, polite, warm, and professional lady who has extensive fieldwork experience in a variety of research areas. Although now based in an urban neighborhood, she grew up in a rural area in KwaZulu-Natal and thus has a deep understanding of traditional Zulu culture. All of these characteristics of Sibongile's identity, personality, and positionality were crucial to my research.

Although Sibongile was my main RA, one or two other experienced RAs (either S'Bosh or Winnie Langa) accompanied us in certain areas, predominantly for safety and transcription reasons. I employed these RAs on the basis of Sibongile's and local colleagues' recommendations. However, as she was my principal RA, I have chosen to focus on Sibongile in this chapter.

Cultural sensitivity is a key ethical principle, especially when conducting research outside your 'home' culture. Central to this is an understanding of cultural beliefs and local traditions and practices. This can then establish a rapport and effective interchange between the receiver and the transmitter of information, which is important for knowledge sharing and transmission to be effective (Moser and Dilling 2007).

The fact that Sibongile and my other RAs shared a common ethnic background and language with the respondents in Zulu-speaking areas, combined with their ability to establish a good rapport with research participants, meant that there were very few positionality concerns regarding the dynamics between the interpreters and research participants.

Sibongile's friendly personality and shared cultural insights and knowledge were pivotal to the formation of effective interpersonal relationships with research participants and their willingness to participate. Her remarkable ability to establish a rapport with research participants was underpinned by her values for mutual respect and cultural norms. She explained: 'it is sometimes difficult to fit in if you don't know the culture, like with the black people, you have to respect the culture, you know you have to respect certain things and other people, so they will respect you too.' Similarly, Meth and McClymont (2009: 919) reflect on Sibongile's critical role as assistant researcher for their research project on men and masculinities, as follows: 'Sibongile's warm and polite personality was absolutely key to the success of this project, and central to the effectiveness of each method.'

Sibongile was not only vitally important in the rapport and relationships with our research participants but also in her relationship with me. On our often long drive home we would reflect on the day's events and interactions and debate the findings. In retrospect, I have realized that these were some of the most crucial periods where I developed my research ideas, focus, and critical interpretations of data. I would document these discussions in my fieldwork diary, which became a very useful information source during my write-up process. Of course, Sibongile's perspectives were shaped by her own personal subjectivities and agenda and were thus often different from my own. This is an important element of the researcher / research assistant relationship – listening to Sibongile's reflections often challenged my own understandings and encouraged me to think more critically about my research.

Looking back, I am confident that Sibongile was absolutely the best choice of RA because of all her positive qualities discussed above. Certainly, such attributes are key considerations for researchers when recruiting a RA.

Effective two-way communication

There is much more to an interpreter's role than the verbatim translation of questions and responses; in addition to overcoming the language barrier, they also act as a kind of cultural broker. This is a two-way process, involving the effective communication of concepts and questions from the researcher to the participants and the accurate understanding and representation of responses by the researcher. Sibongile was invaluable for me in insightfully and sensitively brokering this two-way communication process.

Words can signify different things in different languages and cultures (Temple and Edwards 2002). When preparing my survey and interview schedules, I took it for granted that there would be a direct translation for the word 'climate change' in Zulu with the same meaning. Fortunately, prior to commencing fieldwork, Sibongile explained to me that the terms climate and weather are often used interchangeably in Zulu; if translated literally, climate change can mean 'weather change.' Having realized this, I was able to tailor the wording in my interviews and surveys to accommodate this language technicality (see also Gent in this volume on the need for accurate local understandings of key words and phrases). Sibongile also applied her own knowledge and understanding to ensure effective communication of the research questions:

> After we discussed your work and you explained to me what is climate change when I was doing the work I didn't say climate change exactly. I was explaining in a different way with examples so people understand what you are talking about if you say climate change, but also not to tell them the answers.

Sibongile's comment affirms the importance of effective communication and clarification between the principal researcher and RAs to avoid misinterpretation and misinformation to research participants and consequent inaccurate data collection. Indeed, RAs are 'actively participant intermediaries making judgements which may transform the message received' (Bujra 2006: 175). Effective communication and consequent understanding of key concepts meant that Sibongile could facilitate the research by using terms that were less alien to participants who did not grasp the key research concepts during interviews or surveys.

However, it was not just about the research participants understanding my questions; just as significant was for me to understand their responses. One of my key research findings was the important role that cultural frames and symbols play in shaping people's understanding of and, by extension, responses to climate change. In addition to faith-based beliefs, many respondents, predominantly from the Zulu culture, attributed extreme events, changes in the weather, and climate change to supernatural forces such as divine mythical beings and ancestors. A commonly cited mythical being was the nkanyamba (roughly translated as 'the big snake'), a creature of storm and wind whose powers are linked to destructive tornadoes. While I was familiar with certain relevant cultural beliefs before commencing my research, I would not have grasped fully their meaning and significance without Sibongile willingly sharing her experience of and insights into the Zulu culture with me. She explained:

> When people were telling us about the snake in the sky, ancestors and those others, they were telling us things *they* know. It's *their* understanding, for them, it's what it means. But others might not understand that too.

Moser and Dilling (2007) explicate that, to be successful, communication and interaction about climate change must consider the specifics of the receiver of the information and potential communicative obstacles (e.g., language and cultural differences). Sibongile was key to communication and interaction in the field; she enabled me to reflect analytically and grasp the significance of cultural folklores, 'myths,' and cultural beliefs through sharing her critical cultural perspectives. As such, I fear that without Sibongile and the link she provided to participants from the Zulu culture I would have misinterpreted or never uncovered these important findings.

This issue is of much broader significance to many research projects where RAs provide critical insights into local cultural beliefs that are often far-removed from the personal worldviews of foreign researchers, yet are key to research outcomes. A significant related issue that I grappled with was the communication of such unique terms and concepts into academic language for my thesis. In attempting to make an ethical choice in respecting participant's worldviews or beliefs I chose to refer to original terms in italics and provide an explanation for each of these when first introduced (again see Gent in this volume for further discussion on translation and representation).

In her role of translating and interpreting, Sibongile was not simply doing a job with passive detachment; in fact acting as a linguistic and cultural broker had quite an impact on her. During post-fieldwork discussions Sibongile noted the widespread belief in the supernatural powers of various mythical beings as our main research finding. Reflecting on her research experience she wrote,

> Your research was interesting because it was make me to rethink more what my granny told me while I was young about the big snake if there is a big storm in my area. Now I can tell other people more about climate change too.

Here, and on several other occasions, Sibongile emphasized the educational and empowerment benefits she experienced from being a part of the research. These reflections reveal the significance of accounting for and maximizing the potential benefits for a RA beyond financial gains.

In addition to the effective interpretation and translation of participant responses, Sibongile was sensitive to the way in which things were said or, in some cases, to what was not said. Her experience in using a variety of research tools in diverse contexts has enhanced her sensitivities toward skewed answers and she was able to read 'between the lines' of responses. One issue that she sensed in this respect was the anticipation of benefits following my research. A survey respondent noted when speaking to Sibongile in my absence, 'I wish this survey could do some change to our nation and develop our standards of living.' On another occasion she explained to me:

> When you were with me today at the settlement they changed what they were saying; some told the lies because they thought you were going to maybe help them with electricity or something. Sometimes they think white people have money, even if you explain it differently. Or maybe some think you are from the government because often they get them there.

Together with my reflexive analysis, Sibongile's critical reflections on these and other encounters were crucial to gaining a deeper understanding of participant responses. Moreover, Sibongile's ethical discretion in dealing with these suspected mistruths was crucial to the research process. For Sibongile (and myself), the ethical resolution was not to confront or challenge such responses, as this might have created tensions and weakened relational ties, but rather to ask probing questions and invite elaboration. This type of probing often led to different and detailed responses. As Sibongile noted,

> I know how to approach people, it is a skill you learn to work with people, to talk to them. If someone is giving answers that I am unsure if it is the truth I know it is important to try ask things differently and also make people feel comfortable.

Emotional involvement

Principal researchers are responsible for minimizing harm to both research participants and RAs or interpreters (Hennink *et al.* 2011). This applies not only to issues of safety or physical harm but also to emotional harm. Notably, for Sibongile, a key issue was the feelings of care and responsibility she felt toward research participants who had raised expectations about the research outcomes (see Staddon in this volume for a detailed discussion on participant expectations and 'giving back'). This took an emotional toll on her:

> When we come there they look to us for help, 'cause people are desperate. They looking for development and think maybe we are coming to help them with houses and basic needs. Sometimes that thinking makes me feel not nice to do this kind of work.

This was one of several occasions when Sibongile expressed emotional angst about the wellbeing of research participants. Therefore, clarification of my role as a researcher was important not only for dealing with participant expectations in an ethical manner but also for supporting Sibongile's emotional experience in the field and the ethical responsibilities she felt. As she clarified, 'but once you have explained what you are doing there then I feel more comfortable in myself to do the work.' In addition, Sibongile expressed emotional benefit from sharing research information with interested participants as follows:

> If people didn't know we briefed them after the survey on what climate change is and what the causes are. So I think we were able to give them some education through this research and that is important to me; they gained something. You can't just go into a community, take their information and leave – that is not right.

Recognizing this emotional benefit for Sibongile, I encouraged her to take the time to do this 'briefing' with interested participants.

There are many ways in which research assistants can be emotionally affected or harmed during the research process, especially when conducting research on sensitive issues such as domestic violence or health. In a rare overt reflection, Meth (2003: 155) explains that her research assistant's diarized feelings of stress and angst resulting from participant's shared experiences of domestic violence raised critical questions for her about how 'academics may neglect or overlook the emotional costs of research on assistants and interpreters.' She thus emphasizes the importance of communicating with assistant researchers about their experiences of and views on the research process and of offering support where possible (Meth 2003).

Acknowledging the centrality of research assistants and interpreters

In the above sections I have attempted to include Sibongile's voice and emphasize her role as an active agent shaping and contributing greatly to my research process. I have also reflected on how we negotiated some of the complex ethical and emotional issues that we faced during fieldwork. However, since completing my research I have realized that Sibongile and my other RAs were present more in the background of my thesis, rather than as visible and active stakeholders. This issue is widely applicable to many other theses and other academic research sources.

In short, greater emphasis needs to be placed on the explicit acknowledgement and critical representation of assistant researchers as visible and central actors in the research process. There are several ways to facilitate the effective representation of RAs' voices and experiences. For instance, Meth (2003, 2009) demonstrates the value of encouraging research participants to regularly diarize their perspectives on the research, which can then be referred to and included in the research write-up. Capturing research participants' viewpoints on the research in recorded interviews at various stages of the research process has also proved effective (Temple and Edwards 2002). While the literature on these issues is underdeveloped, there are various exceptional examples to learn from where academics have made their research assistants 'visible' in their research writing by acknowledging their centrality to the research and including their viewpoints (e.g., Temple and Edwards 2002; Meth 2003, 2009; Loftus 2005).

There are many additional pertinent ethical issues and dynamics of working with interpreters or RAs, such as power relations between researcher and research assistant (e.g., regarding monetary payment) and gender issues, that I have not addressed in this chapter. Further, there is no 'secret recipe': the way in which ethical dilemmas emerge and are addressed is shaped by personal judgments, values, and researcher / researcher assistant / participant relationships. These relationships are defined by various subjective, cultural, historical, institutional, and other important elements that define our social positions in the world. Therefore, my aim in this chapter has not been to prescribe what comprises ethical behavior when working with interpreters or RAs, but to encourage researchers to be sensitive to the multiple ethical dimensions of these relationships and to acknowledge explicitly the key role of RAs in the research process.

In a post-fieldwork interview, I asked Sibongile if she realized how important her contribution to my research was and that I could not have completed it without her. She replied: 'No, I didn't realize that, how important I maybe am. Nobody has asked me that before so I didn't think about it.' Sibongile's response was upsetting for me, as I realized that I had paid inadequate attention to the ethics of Sibongile's emotional experience of the research and had not adequately emphasized her value throughout the research cycle and to my final thesis product. In retrospect, although I recognized the critical role played by Sibongile, as reflected in the opening excerpt to this chapter, I am acutely aware that I could have done more to include Sibongile in data analysis and incorporate her perspectives in my final thesis. It is my hope that in some way, this chapter has put some focus and recognition on Sibongile and the critical role of assistant researchers more broadly and will encourage others to learn for the future.

Acknowledgments

This chapter is dedicated to Sibongile, to whom I owe much gratitude for her invaluable input to my research and to this chapter, as well as for being such a dedicated and supportive fieldwork accomplice.

Recommended readings

Meth, P. with Malaza, K. (2003) 'Violent research: The ethics and emotions of doing research with women in South Africa,' *Ethics, Place and Environment*, 6 (2): 143–159.

This paper is particularly relevant for considering and preparing for the emotional costs of research on assistants and interpreters. Drawing on the example of her assistant researcher's emotional experience of her research project, Meth emphasizes the importance of communicating with assistant researchers about their experiences and views on the research process and offering support where possible.

Meth, P. and McClymont, K. (2009) 'Researching men: The politics and possibilities of a qualitative mixed methods approach,' *Social and Cultural Geography*, 10 (8): 909–925.

This paper considers ethical concerns and questions about fieldwork with a particular focus on the challenges and opportunities of a mixed-methods approach, specifically in researching men. The paper includes an explicit reflection on the key role of the research assistant for the success of the research project and effectiveness of each method used.

Temple, B. and Edwards, R. (2002) 'Interpreters/translators and cross-language research: Reflexivity and border crossings,' *International Journal of Qualitative Methods*, 1 (2): 1–12.

This article argues that recent calls for reflexivity in qualitative research should more explicitly include cross-language research with interpreters. Drawing on two research projects as illustrative examples and extending the concept of 'border-crossing,' the authors develop an insightful argument for recognizing the active role of interpreters in producing research accounts and the importance of making the interpreter visible in research.

Note

1 A traditional leader or chief headman who presides over tribal land under traditional authority.

References

Apentiik, C. R. A. and Parpart, J. L. (2006) 'Working in different cultures: Issues of race, ethnicity and identity', in V. Desai and R. Potter (eds), *Doing development research*, London: Sage.

Bujra, J. (2006) 'Lost in translation? The use of interpreters in fieldwork', in V. Desai and R. Potter (eds), *Doing development research*, London: Sage.

Hennink, M., Hutter, I. and Bailey, A. (2011) *Qualitative research methods*, London: Sage.

Loftus, A. J. (2005) *A political ecology of water struggles in South Africa*, unpublished thesis, Oxford University.

Meth, P. with Malaza, K. (2003) 'Violent research: The ethics and emotions of doing research with women in South Africa,' *Ethics, Place and Environment*, 6 (2): 143–159.

Meth, P. and McClymont, K. (2009) 'Researching men: The politics and possibilities of a qualitative mixed methods approach,' *Social and Cultural Geography*, 10 (8): 909–925.

Moser, S. and Dilling, L. (2007) 'Toward the social tipping point: Creating a climate for change,' in S. Moser and L. Dilling (eds), *Creating a climate for change: Communicating climate change and facilitating social change*, Cambridge: Cambridge University Press.

Temple, B. and Edwards, R. (2002) 'Interpreters/translators and cross-language research: Reflexivity and border crossings,' *International Journal of Qualitative Methods*, 1 (2): 1–12.

7 Doing it together: ethical dimensions of accompanied fieldwork

Experiences from fieldwork in India

Jenny Lunn and Alessandro Moscuzza

> Working in the field together is excitement, a spice to life, the best form of wild ride … there's nothing like hitting the road with your husband or wife or partner with children along.
>
> (Stars *et al.* 2001: 76)

There is a predominant image within development studies of the solitary researcher heading off alone to far-flung foreign places. It is assumed that an independent researcher will, by their very nature, be a detached and objective observer and that, by implication, their research will be more rigorous (Gottlieb 1995). The reality, though, is that accompanied fieldwork is commonplace. Some people go with their spouse or partner while others take their child or children; for some researchers fieldwork is a whole family experience while others are visited in the field by siblings, parents, or friends. Equally common is being accompanied by fellow researchers and colleagues, but this chapter focuses on family members.

Jenny

Prior to my departure on doctoral fieldwork, I had never seen any reference in my reading of development fieldwork literature to being accompanied. I simply assumed that I would go alone because no one and nothing had indicated otherwise. With the fieldwork departure date on the horizon, I was emotionally preparing myself for a long separation. My long-term partner, Alessandro ('Alex'), would probably come to visit once, taking his summer vacation in my field location, but – realistically – neither of us thought our relationship would survive the one-year separation. With sadness we began wrapping up the relationship both emotionally and logistically. I was excited at the prospect of my field year but sad that it came at the expense of my long-term relationship.

One day it suddenly struck me that effectively nothing was preventing Alex from coming with me. Although I had not come across any examples in the literature, neither had I seen any regulations prohibiting it. I checked with my supervisor whether accompanied fieldwork was acceptable and, receiving a positive response, invited Alex to join me. This chapter explores this decision and its implications.

Most researchers mention the support of family members in the acknowledgements of their thesis or academic publications, but few have reflected critically in their publications on the implications of the presence of family members in the field and on the research process. Most of the existing literature is found within social and cultural anthropology, where long periods of ethnographic fieldwork are a more common practice. More recently, as reflexive writing has spread across the social sciences and beyond, there have been contributions from researchers in other disciplines such as geography, geosciences, development studies, sociology, and tourism studies. While most of the references cited in this chapter relate to fieldwork in the Global South, there are also a few examples from fieldwork conducted in difficult or remote environments of the Global North. Drawing from our experiences and the literature, in this chapter we offer some thoughts and reflections to others in a similar situation.

The existing literature on experiences of being accompanied on fieldwork can be divided into three main types according to which family members are present: adults, children, and families. First, there are accounts which mention the presence of spouses, siblings, or partners in the field, including cases of getting married during the period of doctoral fieldwork (e.g., Missenden Centre 2006; Shepherd 2007) or marrying someone from the field location, whether a local person (Gordon 1998; Sinclair 1998) or key informant (Irwin 2006). Second, a few female researchers have written about being with their offspring of various ages in the field, sometimes with spouses or parents dropping in to assist with child care (e.g., Young-Leslie 1998; Cupples and Kindon 2003; Sciencewoman 2007; Gottlieb 2008; Ramage Macdonald and Sullivan 2008). Third, some researchers take their spouses and children because, as a family, there is no question of being separated. In some cases, both parents are academic researchers (e.g., Counts and Counts 1998; Young-Leslie 1998; Starrs *et al.* 2001; Shepherd 2007; Goldstein 2008); in other cases just one parent (e.g., McGrath 1998; Lorimer and Spedding 2005). On the other hand, some families split up for fieldwork, generally with older children staying at home and younger children in the field with attendant practical, logistical, and emotional problems (e.g., Counts and Counts 1998; Goodenough 1998).

It is important to note that writing about accompanied fieldwork is not the sole domain of the researcher. Spouses and partners have shared their experiences, some in an academic format (e.g., Goodenough 1998) and others in the form of memoirs (see the list in Gottlieb 1995), while some children have also been given the chance to recount their experiences in an academic article alongside their parents' reflections (e.g., Petersen *et al.* 1998; Starrs *et al.* 2001). Meanwhile Lorimer and Spedding (2005) examine the highly entertaining logbook of a family fieldtrip with written and pictorial contributions from the various participants.

While memoirs of family field experiences are interesting, of particular concern in the context of this book is how the presence of additional people in the field raises particular ethical questions such as: Who decides who goes to the field? What are the consequences of taking or not taking family members to the field? How does the presence of family members affect the identity of the researcher and the perceptions of the research participants? What should be the role of those accompanying in the field? And how does the presence of family members affect the data collection process? In this chapter we explore these questions through the case of our time spent in India.

The power to decide

A fundamental ethical question is who decides who should and should not be present during fieldwork. It may seem logical that the researcher has the power to decide, not only because it is their project that will be affected by the accompanied status but also because they may have some understanding of the research context and, in practical terms, its suitability for accompanying family members. However, is it right that one person makes the decision of whom to include and whom to exclude? Should family members be presented with the options and asked to choose? What if a family member invites themselves along: does the researcher have the authority to deny their presence if it is not appropriate?

While adults have more freedom to accept or reject an invitation to the field, children may have the decision imposed on them. In the accounts of fieldtrips with children cited in the previous section, there is little mention of children being given a choice, apart from Counts and Counts (1998), who took their children when they were young but gave them the choice when older. While those deciding to take children believed that the field experience would be good for them (e.g., McGrath 1998; Cupples and Kindon 2003), it would be interesting to read some retrospective reflections from both the parents and children years or decades later as to whether it was actually beneficial for them in the short and long term.

A second ethical question concerns the appropriate compensation for family members who are leaving their home setting and may be making sacrifices to follow the researcher. Children, for example, may be withdrawn from school, and suitable alternatives need to be found in the field location. Spouses and partners, on the other hand, may rely on an understanding employer to grant extended leave or may need to resign from their job. In both cases, the interruption of education or career could have long-term implications. Doing fieldwork together entails putting the researcher's needs and interests first, and this prioritization needs to be sensitively brokered and mutually agreed by all family members involved.

Conversely, it is also important to consider the practical, emotional, and ethical implications of choosing to leave family members behind. In the case of children, in particular, this raises further ethical issues about responsibility for their care and welfare. Counts and Counts's (1998) 13-year-old son decided to stay at home on later fieldtrips in the care of his 'long-suffering grandparents' in order to avoid missing a year of school. Meanwhile, during the later fieldtrips of the Goodenough (1998) family, the younger children went to the field while the older children stayed at home to attend college; one of those at home struggled without parental support and needed extensive counselling. On the other hand,

Gilmore (1998) took her daughter to Tahiti but she was extremely homesick and the fieldtrip had to be curtailed.

As already mentioned, prior to fieldwork we had assumed that Jenny would go alone until a late change of mind.

Alex

I received Jenny's invitation to join her on her doctoral fieldwork year with mixed feelings. On the one hand I was excited at the sudden prospect of leaving behind my office job in London and moving overseas to do work in a developing country, which is what I had studied for. On the other hand this was tempered by some doubts concerning the uncertain prospect of finding something meaningful to do in India and the potential long-term impacts of having a break from my job at a time when I was consolidating my position and moving up the career ladder. There was also the daunting prospect of negotiating the bureaucracy involved in getting the right kind of visa and various other practicalities.

Once my employer agreed in principle to a one-year leave of absence, it was decision time. Jenny and I discussed further the implications of me going, and I set out my conditions: principally for me to make this a meaningful and not a wasted year by having my own professional opportunities in India, while also providing support for her fieldwork. I was taking a risk but, having weighed the pros and cons, I decided to accept the offer. Ultimately the final decision was mine, albeit made after extensive discussions with Jenny, my family, and my colleagues. No doubt the decision-making process would have been different had we been married with children, in which case there would have been more complex ethical questions and practical challenges to face.

It was thus with great excitement that we set about the last few months of planning and preparation for our trip to India, which was now a shared experience. We went to weekly language classes together, had our vaccinations, booked our flights, packed our belongings into storage, and said our goodbyes to family and friends. An exciting adventure lay ahead.

Perceptions and positionality

Every researcher has a range of personal and social identities, ranging from ethnicity and age to gender and religion. Whether a field researcher is accompanied or not, and by whom, constitutes one of these identities. The ethical dimension of this concerns the way in which researcher identities are expressed in the field setting and perceived by the people there, as well as the ways in which these positionalities can affect the research process (see also the chapters by Dam and Lunn, Godbole, and Kovàcs and Bose in this volume for other examples of negotiating positionality in the field).

Being accompanied or unaccompanied may affect whether a researcher conforms with or contradicts socio-cultural norms in the fieldwork location. For example, in societies where a woman in her thirties would be expected to be married, going to the field accompanied by your spouse may lend respectability and acceptance, whereas leaving your partner at home (for whatever well-justified reason) may cause people to make negative judgments and create barriers to building rapport (e.g., Cipollari 2010).

Some of the most interesting accounts about the impact of accompanied status on positionality in the field have been written by people who have conducted successive field visits and compared their experiences (e.g., Flinn 1998; Linnekin 1998; Twyman *et al.* 1999; Latvala 2001). Jenny could also compare experiences because she had carried out fieldwork alone in the same place ten years previously.

Jenny

When I conducted research for my master's degree dissertation I was in my early twenties and alone in the field, as Alex was simultaneously doing his own field research on the other side of the world. I felt that interviewees, particularly middle-aged males, tended to regard me as slightly inferior because of my age and single status, although I was still granted access, probably because of my ethnicity and nationality (see Dam and Lunn in this volume for further reflections).

By the time of my doctoral fieldwork I was in my early thirties and accompanied by Alex. Although we were not married, we pretended to be so while in India because it would have been culturally unacceptable for us to be living together. When meeting research participants I often mentioned 'my husband' in the course of conversation; my finger ring was also a visible identifier. It is difficult to say whether this was a subtle yet deliberate way of presenting myself as more mature and respectable or just a natural reflection of my identity.

However, being accompanied, 'married,' and a little older, I felt that interviewees of all ages and genders respected me and I was treated more seriously as a professional researcher than on my previous visit. I cannot claim that accompanied status was the sole or main reason for being treated more seriously; neither would I wish to claim that lone researchers would be considered less seriously, but I think it certainly played a role in my experience.

In many societies of the Global South a woman of my age would almost invariably have children, and I was often asked about this. Among the people that I interacted with, it was those from poorer communities who were most shocked that I did not have any children at my age. In their minds the gods must have looked unfavorably on me not to have granted me a full brood of infants by my thirties (see also the chapters by Godbole and Staddon in this volume for similar reactions). Many times I had to explain that, in countries such as the UK, some women choose to study at school and university for longer, get married older and have children later or even not at all. Time and again, women said to me: 'Next time you come to visit, you will come with a baby!' Were I to do that in the future my status would no doubt be one of much closer acceptance as 'one of them' (see Twyman *et al.* 1999 for a similar example).

Overall, I feel that being accompanied by a partner was beneficial for my fieldwork in terms of cultivating a sense of respectability, while being unaccompanied by children colored some people's perceptions of me, though not to the detriment of carrying out my research. However, if I had had children it might have been perceived as inappropriate that I was working and not attending to domestic duties.

I suspect that accompanied status is much more significant for female researchers. Alex did not feel that his marital status affected people's perceptions of him, although introducing me to colleagues and business associates as his 'wife' no doubt added a layer of respectability to his persona. Rather more significant in presenting himself to people were his academic qualifications and professional affiliations, which carry weight in the Indian context and many other countries and cultures.

Responsibilities and roles

While a period of fieldwork may be dynamic, exciting, and fulfilling for the researcher, the accompanying family members may not share the same enthusiasm for the research topic or for being in the particular field location. So having invited, persuaded, or cajoled family members to the field, should the researcher take responsibility to ensure that their presence is valued, their time is used productively, and they are happily occupied, particularly if they have made sacrifices in order to be there? Or does the researcher already face enough responsibility in the field to deliver their project on time without having these additional pressures? In the case of adults, should it be up to the individual to make the most of whatever situation they find themselves in? Since the activities of accompanying children will vary according to their age and the duration of the fieldwork period, this section focuses solely on the roles of spouses and partners.

There are various possible roles for an accompanying adult, including a research assistant in the fieldwork, a professional with their own work and priorities, a domestic support taking responsibility for household duties and child care, or a 'year off' and a chance for a holiday and travelling; they may even play a combination of these roles. Whichever role an accompanying partner takes there are ethical dilemmas relating to who has the power to decide about the role and its effect on the research project. For example, involving an accompanying partner as a research assistant comes with its own set of ethical dilemmas concerning leadership, payment, bias, and other issues (see Le Masson in this volume for further discussion of this issue). Alternatively, if the accompanying partner is predominantly providing the domestic support, how might this ancillary role make them feel (see Lorimer and Spedding 2005), particularly if they have sacrificed their job to be in the field? Scheyvens and Nowak (2003) suggest discussing the accompanying spouse or partner's role before departing on the fieldtrip in order to be clear about roles and expectations. That is certainly what we did.

Alex

As already mentioned, I accepted Jenny's invitation to accompany her in India on the condition that, while being available to help her as required, I would also find some work of my own that would contribute to my career development. Thus my two distinct roles were very clear from the start. In principle they were equally important; in practice I spent a greater proportion of my time on my work, although that was not to the detriment of supporting Jenny. Overall I feel that there were three main ways in which I supported Jenny during our year in India.

First, in practical terms I took responsibility for most of the logistics of moving to another country, such as looking for an apartment to rent and dealing with property agents; making practical arrangements such as setting up mobile phone contracts and an internet connection; and establishing contacts with a local doctor and dentist for our welfare. In part this was to allow Jenny to get started with her fieldwork as soon as possible; it was also a deliberately gendered division of labor, as such dealings in the public realm are more commonly carried out by men in India.

Second, in emotional terms I provided encouragement, support, and advice to Jenny on a daily basis. However, it is important to highlight that this type of support was two-way. Living and working in a different culture can be difficult and, at times, frustrating, stressful, and emotional. Unlike Jenny, I had never previously been to India so I found it beneficial to come home at the end of the day to someone with whom I shared a cultural background and could discuss the various challenges that I faced both within the workplace and in the city.

Third, in terms of the actual field research I supported Jenny in a variety of ways. On her request, I accompanied her to some interviews and on field visits where she might have felt unsafe or vulnerable on her own as a female. I was also able to assist with her data collection (see the next section). Furthermore, in what could be considered a supervisory role, I asked Jenny regularly for progress updates on her ongoing data collection and helped her to design realistic schedules for her remaining activities so that the research would be completed within the timeframe. Although we had met at university while doing the same master's degree in development studies, our academic and professional interests had since diverged, and Jenny's research topic was not of particular interest to me. Yet our shared academic background and training meant that I was still able to understand her research and offer constructive advice that might not have been possible had I been a professional in a completely unrelated field (again see Le Masson in this volume for a similar situation).

Jenny was grateful for my support in each of these three spheres but, as mentioned earlier, it was equally important to me that I found employment in an area that was closely related to my interests and which would help with my professional development in the long term. Fortunately, through both UK and Indian contacts, I was able to find work with an institutional body which complemented my experience of working for a UK government department. In addition, through Jenny's pre-existing networks in Kolkata, I was able to get some freelance work with three different NGOs. These experiences definitely enriched my work experience and helped me to develop my professional profile. Had I been unable to find any employment, or else been engaged in work that was unstimulating or irrelevant to my career development, then I would have felt unsatisfied and that I was 'wasting' the year. This frustration might also have affected my ability to offer quality support and advice to Jenny.

All in all, I feel that my having a dual role in the field was beneficial for both of us. There was also a certain amount of interesting cross-over, as Jenny's contacts resulted in work for me and my contacts were of use to Jenny's research. To other partners considering going on fieldwork, I would strongly recommend that expectations and roles are discussed thoroughly in advance, although some flexibility will no doubt be required once in the field. I suggest that the key elements are mutual agreement, personal benefit, independence, and a sense of purpose, although this may vary according to personal characteristics.

Facilitating data collection

As already discussed, being accompanied during fieldwork creates particular positionalities in relation to the setting and research participants. Since this can affect the collection of data, which is the core purpose of the fieldwork, there are certain ethical questions concerning whether the presence of family members helps or hinders the collection of data, in what ways, and with what implications (see also the chapters by Le Masson and Taylor in this volume on the use of a partner as a research assistant and family members as gatekeepers respectively).

In the literature, the only mentions of negative impacts of accompanying family members on data collection come from scientists and physical geographers rather than from social scientists, and these are mostly practical concerns. For example, the Murray family's fieldtrip was significantly shaped by the feeding times of the accompanying three-month old baby (Lorimer and Spedding 2005), while one geoscientist struggled to reach some field sites while wearing her baby in a papoose (Sciencewoman 2007). Most of the examples in the literature, however, demonstrate the positive impacts of family members on data collection, and broadly divide into two types.

First, being accompanied can facilitate access to the research setting by helping in the development of a rapport with respondents. In this respect, the impact of babies and young children are mentioned most in the literature, particularly in the case of ethnographic fieldwork. Children naturally break the ice by innocently making contact, establishing relationships, and paving the way for adult interaction (e.g., Counts and Counts 1998, Shepherd 2007, Scanlon Lyons 2008). Researchers who have married a local person during the fieldwork, as in the case of Gordon (1998) and Irwin (2006), no doubt gained extra access and insight into the research community, although both faced difficulties regarding their status and positionality.

Second, the accompanying person can enable the data collection and analysis to be more effective. In two cases in the literature, the female researcher's subject was motherhood, and the fact that they had their own children with them not only influenced how respondents perceived them (i.e., they understood what being a mother is like) but also gave them an insight into the daily struggles of having a child in that particular field location (e.g., Latvala 2001; Cupples and Kindon 2003). In other cases, the accompanying spouse was an insider to the community being researched and could thereby facilitate access and ensure correct interpretation of what the researcher was observing and hearing. Roald (2001), for example, as a Western convert to Islam researching women's experience of Islam, felt that her Palestinian husband's presence increased the effectiveness of interviews. Meanwhile, Sinclair (1998) was forced to re-evaluate her research and some of the key issues in the light of her daughter's reflections on what she observed in the field. Similarly, the innocent observations of McGrath's (1998) children led to insights into the local culture that the adults missed.

While developing rapport, facilitating access, and the interpretation of responses is important for all researchers, the ethical dilemmas lie in whether family members should be strategically – maybe unwittingly – used by the researcher. Is it right to use someone's presence or position without their explicit consent? In our case, Jenny sometimes requested Alex's assistance and sometimes it was initiated by him, and although she utilized his presence and position for the advantage of her fieldwork, it was with his consent and was in no way manipulative. In all the cases of his involvement it was also positive and productive for the data collection.

Alex

I think that I influenced Jenny's data collection in two main ways. First, through my own work I identified a number of useful contacts for Jenny and built rapport with these people. Personal introductions are important in the Indian context, so my prior contact was very useful when Jenny requested an interview. Since I was working in an institutional context I opened up contacts in this sector that Jenny might not otherwise have obtained.

Second, as mentioned in the previous section, I sometimes accompanied Jenny to interviews for reasons of safety, and my presence often helped to break the ice through informal conversation before the interview started. Interviewees were interested in my background and profession. In the case of male interviewees, the presence of another male in the room may have helped them to feel more relaxed and resulted in a more free-flowing conversation. We were also able to discuss the content of these encounters afterwards, which Jenny found very useful in terms of checking her interpretation of responses (see also the chapters by Le Masson and Leck in this volume on triangulating responses with research assistants).

Jenny could no doubt have carried out the data collection alone and unaccompanied. However, I feel that my presence and the resultant positionality this created helped her to collect more and better data through opening up a wider range of interviewees, creating a rapport with respondents, and being a 'second ear' in interviews. I did not mind my physical presence, academic input, or local contacts being used for the good of her project. However, other partners in the field may not necessarily have the same kind of experience, so I would recommend that researchers carefully analyze the impact of their accompanied status on their positionality in the field setting.

Together or not?

This discussion has reflected on some of the ethical questions surrounding accompanied fieldwork, particularly its impact on the research process and outcomes, but perhaps the final word should be about the impact of accompanied fieldwork on the relationships of those involved. We left for India with a positive attitude. Like the McGrath (1998) family in Tonga, we saw it as an opportunity to share an adventure together. In fact, we both had one of the best years of our lives and that was because we experienced it together. The head of the Starrs family concluded that their fieldwork together was exciting and unforgettable:

> I can think of only one thing in my life that competes with the rewards and pleasure of fieldwork – short-term or long-term – and that is time with family. To bring fieldwork and family together is a joy, if a refined and nuanced one. The reciprocal of 'worry' is, after all, nothing less than thrills, and when with family, those are all shared.
>
> (Starrs *et al.* 2001: 76)

The available literature invariably concludes that there are both advantages and disadvantages of being accompanied in the field. For us, the positives far outweighed the negatives and with the sacrifices came tremendous opportunities. Alex's presence as an accompanying spouse certainly affected Jenny's positionality and served to foster a sense of respectability. Alex made an invaluable contribution to Jenny's fieldwork in intellectual, practical, and emotional ways. This was helped by the fact that we had a shared academic and professional background. Alex took a risk in taking a year's leave from work but it paid off. He found relevant employment in India that was satisfying and that contributed to his career development. There was also valuable cross-over with Jenny's contacts being useful for Alex and vice versa.

The power to decide was vested in Alex, the invitee to the fieldtrip, who, as an adult, was freely able to accept or decline. Other ethical questions were negotiated by good communication of intentions and desires from the outset, a clear setting out of roles and expectations in the planning stage, and a combination of interaction and independence once in the field.

It can be difficult to separate academic research from personal life in the field, even more so when accompanied by family members (Cupples and Kindon 2003; Lorimer and Spedding 2005). For us, being in the field together has become another chapter in our journey through life together and strengthened our relationship.

Not all researchers will be able to manage the logistics of accompanied field-work, as there are costs, careers, and possibly children to consider. Nor will all those who are accompanied in the field have such a positive experience as ours. Neither has this chapter had the scope to reflect on same-sex partners, whose presence in the field is likely to have a different set of issues and implications. However, we hope that this chapter adds to the limited literature on accompanied fieldwork that we wish we had discovered prior to our own experiences.

Recommended reading

Cipollari, C. (2010) 'Single or married? Positioning the anthropologist in tourism research,' *Journal of Tourism Consumption and Practice*, 2 (2): 30–58.

A reflexive account that considers how marital status affected the researcher's positionality in the field. Highlights the differences between what the researcher thinks are the most significant parts of their identity compared with how the host community perceived them and what behavior was expected.

Cupples, J. and Kindon, S. (2003) 'Far from being "home alone": The dynamics of accompanied fieldwork,' *Singapore Journal of Tropical Geography*, 24 (2): 211–228.

Recounts experiences of being accompanied in the field – by children in the case of Cupples and by other female researchers in the case of Kindon. Explores how the people with you impact and enrich the research, and calls for a removal of the distinction between research and self.

Flinn, J., Marshall, L. B. and Armstrong, J. (eds) (1998) *Fieldwork and families: Constructing new models for ethnographic research*, Honolulu: University of Hawaii Press.

A compilation which brings together varied experiences of taking family members on ethnographic fieldwork expeditions. Although the examples are all drawn from fieldwork in the Pacific region, the content is relevant to any location; this is highly recommended for anyone considering taking family members to the field.

Lorimer, H. and Spedding, N. (2005) 'Locating field science: A geographical family expedition to Glen Roy, Scotland,' *The British Journal for the History of Science*, 38 (1): 13–33.

A very readable paper which looks at the Murray family's holiday-cum-expedition in 1952. Reflects on the practicalities of fieldwork accompanied by a three-month old baby and reveals some of the creative content of the entertaining logbook to show how the presence of the family in the field changed the way in which the field location was experienced and understood.

References

Cipollari, C. (2010) 'Single or married? Positioning the anthropologist in tourism research,' *Journal of Tourism Consumption and Practice*, 2 (2): 30–58.

Counts, D. R. and Counts, D. A. (1998) 'Fictive families in the field,' in J. Flinn, L. B. Marshall and J. Armstrong (eds), *Fieldwork and families: Constructing new models for ethnographic research*, Honolulu: University of Hawaii Press.

Cupples, J. and Kindon, S. (2003) 'Far from being "home alone": The dynamics of accompanied fieldwork,' *Singapore Journal of Tropical Geography*, 24 (2): 211–228.

Flinn, J. (1998) 'Single woman, married woman, mother, or me? Defining family and identity in the field,' in J. Flinn, L. B. Marshall and J. Armstrong (eds), *Fieldwork and families: Constructing new models for ethnographic research*, Honolulu: University of Hawaii Press.

Gilmore, S. S. (1998) 'Both ways through the looking glass: The accompanied ethnographer as repositioned other,' in J. Flinn, L. B. Marshall and J. Armstrong (eds), *Fieldwork and families: Constructing new models for ethnographic research*, Honolulu: University of Hawaii Press.

Goldstein, D. (2008) '*Mothering (in) the field:* A roundtable discussion,' panel session at American Anthropological Association Annual Meeting, San Francisco, November 2008.

Goodenough, R. G. (1998) 'Fieldwork and a family: Perspectives over time,' in J. Flinn, L. B. Marshall and J. Armstrong (eds), *Fieldwork and families: Constructing new models for ethnographic research*, Honolulu: University of Hawaii Press.

Gordon, T. (1998) 'Border-crossing in Tonga: Marriage in the field,' in J. Flinn, L. B. Marshall and J. Armstrong (eds), *Fieldwork and families: Constructing new models for ethnographic research*, Honolulu: University of Hawaii Press.

Gottlieb, A. (1995) 'Beyond the lonely anthropologist: Collaboration in research and writing,' *American Anthropologist*, 97 (1): 21–26.

Gottlieb, A. (2008) '*Mothering (in) the field:* A roundtable discussion,' panel session at American Anthropological Association Annual Meeting, San Francisco, November 2008.

Irwin, K. (2006) 'Into the dark heart of ethnography: The lived ethics and inequality of intimate field relationships,' *Qualitative Sociology*, 29 (2): 155–175.

Latvala, J. (2001) 'Reflections from a contact/conflict zone: An analysis in four acts,' *Elore*, 8 (1), available at: www.elore.fi/arkisto/1_01/jlat101.html (accessed 10 January 2010).

Linnekin, J. (1998) 'Family and other uncontrollables: Impression management in accompanied fieldwork,' in J. Flinn, L. B. Marshall and J. Armstrong (eds), *Fieldwork and families: Constructing new models for ethnographic research*, Honolulu: University of Hawaii Press.

Lorimer, H. and Spedding, N. (2005) 'Locating field science: A geographical family expedition to Glen Roy, Scotland,' *The British Journal for the History of Science*, 38 (1): 13–33.

McGrath, B. B. (1998) 'Through the eyes of a child: A gaze more pure?,' in J. Flinn, L. B. Marshall and J. Armstrong (eds), *Fieldwork and families: Constructing new models for ethnographic research*, Honolulu: University of Hawaii Press.

Missenden Centre (2006) *Carole's PhD Diary 2002–2005* (anonymous diary of a geography PhD student), Great Missenden, Bucks: The Missenden Centre for the Development of Higher Education.

Petersen, G., Garcia, V. and Petersen, G. (1998) 'Field and family on Pohnpei, Micronesia,' in J. Flinn, L. B. Marshall and J. Armstrong (eds), *Fieldwork and families: Constructing new models for ethnographic research*, Honolulu: University of Hawaii Press.

Ramage Macdonald, J. and Sullivan, M. E. (2008) 'Mothers in the field,' *The Chronicle of Higher Education*, 24 October, available at: http://chronicle.com/article/Mothers-in-the-Field/45801 (accessed 10 January 2010).

Roald, A. S. (2001) *Women in Islam: The Western experience*, New York: Routledge.

Scanlan Lyons, C. M. (2008) '*Mothering (in) the field:* A roundtable discussion,' panel session at American Anthropological Association Annual Meeting, San Francisco, November 2008.

Scheyvens, H. and Nowak, B. (2003) 'Families and partners in the field,' in R. Scheyvens and D. Storey (eds), *Development fieldwork: A practical guide*, London: Sage.

Sciencewoman (2007) *Bringing baby to the field*, 16 April, available at: http://sciencewoman.blogspot.com/2007/04/bringing-baby-to-field.html (accessed 10 January 2010).

Shepherd, J. (2007) 'Melissa Leach: Village voice,' *Guardian*, 17 July, available at: www.guardian.co.uk/education/2007/jul/17/highereducationprofile.academicexperts (accessed 10 January 2010).

Sinclair, K. (1998) 'Dancing to the music of time: Fieldwork with a husband, a daughter, and a cello,' in J. Flinn, L. B. Marshall and J. Armstrong (eds), *Fieldwork and families: Constructing new models for ethnographic research*, Honolulu: University of Hawaii Press.

Starrs, P. F. with Starrs C. F., Starrs, G. I. and Huntsinger, L. (2001) 'Fieldwork … with family,' *Geographical Review*, 91, vols 1–2: 74–87.

Twyman, C., Morrison, J. and Sporton, D. (1999) 'The final fifth: Autobiography, reflexivity and interpretation in cross-cultural research,' *Area*, 31 (4): 313–325.

Young-Leslie, H. (1998) 'The anthropologist, the mother, and the cross-cultured child: Lessons in the relativity of cultural relativity,' in J. Flinn, L. B. Marshall and J. Armstrong (eds), *Fieldwork and families: Constructing new models for ethnographic research*, Honolulu: University of Hawaii Press.

Part II

Ethical dimensions of researcher identity

8 Revealing and concealing: ethical dilemmas of maneuvering identity in the field

Experiences from researching the relationship between land and rural women in western India

Girija Godbole

'What is your surname?'[1] the son of Parubai, one of my *dalit*[2] respondents, asked me. Clearly he was not pleased with just my first name and wanted to know more, perhaps to place me in the caste hierarchy. 'Godbole,' I replied rather nervously, watching his mother from the corner of my eye. 'So you are a Brahmin!'[3] he commented, glancing at Parubai. I could see a look of surprise on her face, but she quickly composed herself and said, 'Oh that does not matter, she is just like us!' I felt relieved and slightly proud of myself for being able to cross the caste barrier, albeit to a limited extent.

My doctoral research investigated how rural women felt about the changes in their society as a result of the increased incidence of land sale and migration. However, the process of building relationships with village women and eliciting information from them on this subject was significantly shaped by my own identity and positionality in their midst, as illustrated in my account above and as I shall explain in this chapter.

Peak and Trotz (1999, cited in Sultana 2007: 376) suggest that acknowledging one's positionality and subjectivity

> can strengthen our commitment to conduct good research based on building relations of mutual respect and recognition. It does, however, entail abandoning the search for objectivity in favor of critical provisional analysis based on plurality of (temporally and spatially) situated voices and silences.

For me, making a shift toward a more subjective position required some effort, as my graduate training in anthropology in the late 1990s at an Indian university placed an emphasis on maintaining objectivity and value-neutrality while doing fieldwork. Preparing for my doctoral fieldwork focusing on women's views involved reading more on feminist research methods. To adopt a stance closer to feminist methodology meant unlearning the objectivist framework and taking a more subjective, reflexive position. I had to accept that 'the interviewee is not an object, but a subject with agency, history and his or her own idiosyncratic command of a story. Interviewer and interviewee are in partnership and dialogue as they construct memory, meaning and experience together' (Madison 2012: 28).

The literature on feminist methodology also made me more aware of the nuances of power relations between the researcher and the researched. I agree with Maynard (1994) that research becomes a means of sharing information rather than being seen as a source of bias; the personal involvement of the interviewer is an important element in establishing trust and thus good quality information.

In this chapter I explore the significance of different aspects of my identity, particularly the way that they positioned me vis-à-vis my respondents, and some of the ethical dilemmas that I faced in utilizing and manipulating them in the field for the benefit of my research. The first section focuses on the particularly Indian issue of caste; the second section looks at other aspects of my identity including age, marital status, class, and educational background, and how I variously revealed and concealed them; and the third section considers how I positioned myself when discussing sensitive and personal matters with respondents.

Doing research at 'home' in a caste-based society

The area I chose for fieldwork was familiar to me: I was born and brought up in the nearby city, I spoke the same language as my respondents, and I had been visiting some of the villages since my childhood and knew a number of people there. Thus in many ways my fieldwork was at 'home' since it was in a region with which I have familiarity in terms of language, beliefs, culture, and personal networks. There is an implicit assumption that doing research at 'home' or in a familiar place is somehow easier, but I would caution against this. 'Home' is a complex concept and I was well aware of the social distance that existed between me and my respondents in terms of caste, class, education, and other factors.

M. N. Srinivas (1997: 22), the well-known Indian anthropologist, commented that

> when an Indian anthropologist is studying a different caste or other group in India, he is studying someone who is both the Other and also someone with whom he shares a few cultural forms, beliefs, and values. That is, he is studying a self-in-the-Other and not a total Other, for both are members of the same civilization, which is extraordinarily complex, layered and filled with conflicting tendencies.

Thus in the field I was simultaneously an insider, an outsider, both, and neither (Gilbert 1994) (see also the chapters by Dam and Lunn and by Taylor in this volume for similar ambiguities over insider/outsider status).

I come from what could be described as a privileged background: a well-educated, upper caste, financially well-off, socially progressive, urban family from Pune in western India. My parents did not believe in the caste hierarchy and sympathized with the anti-caste movement. They encouraged me to question caste-imposed boundaries and make ties with people from different castes, as well as from other class and religious backgrounds. Although one may determine not to believe in caste hierarchy and try to defy it, however, it is impossible to escape it in India, where it forms the stratification of society.[4] A number of progressive constitutional provisions intended to undo historical inequalities have not been successful in wiping out the undercurrents of the caste system, which can still be felt in a variety of ways. When I decided to do fieldwork at 'home' this inevitably placed me in the social hierarchy, at least in the initial phase.

Traditionally across India, Brahmin, the priestly caste, is considered to be at the peak of the hierarchy. Certainly in rural areas of my home region of Western Maharashtra, many Brahmins used to practice seclusion and treat other castes in an extremely demeaning and exploitative manner. Understandably there is a lot of resentment against them in rural areas (and also in urban areas). I was well aware of this when I decided to study the changing relationship between land and women in rural Maharashtra. The incident described in the opening paragraph was testimony to this, although the young man did not show his disapproval in explicit terms. As someone who condemns this exploitative caste system, I too share this anger and feel that there is nothing to be proud of in being born into the Brahmin caste[5].

Given the caste dynamics of my research setting, I needed to make conscious efforts to bridge any gaps that arose because of my caste; even though caste was not important to me personally, it was significant to many of those around me. For example, in my sample, I made a deliberate effort to include as many *dalit* women respondents as I could. During my interaction with them, I made a point never to turn down any tea or food that was offered. Traditionally Brahmins did not accept food or drinks from people from so-called lower castes or untouchable castes, so this small act had symbolic significance. As per sacred Hindu scriptures, Brahmins are only supposed to consume vegetarian food. The fact that I am a non-vegetarian came as an added advantage in breaking down socio-cultural barriers[6]. To reiterate the fact that I condemned the caste hierarchy I also made comments about how unjust it was for Brahmins to treat the other castes in a demeaning manner, and I often used the term *baman*[7] rather than Brahmin. In our initial interactions, most of the *dalit* respondents found my efforts to transcend the caste-defined boundaries rather amusing. As our contact grew and we had more discussions around caste issues I could see that they were pleased to see an upper-caste urban woman questioning the traditional privileges of her own caste.

Positionality is a critical factor in framing social and professional relationships in the field; it sets the tone of the research, affecting its course and its outcomes (Chacko 2004: 52). It is absolutely essential for researchers working in India to be aware of the complex ways in which the caste system shapes the power relations between themselves and their participants. Respondents considered me, someone from the same region, as an 'insider' and the traditional norms of caste hierarchy became applicable to me. Hence I had to make conscious efforts to overcome the barriers that my 'privileged' caste and class position accorded. By ignoring caste conventions I had an opportunity to demonstrate my dislike toward the caste hierarchy.

How much to reveal?

While my caste identity was shown by my surname, other aspects of my identity such as my age, marital status, class, and educational background were not immediately obvious. My decision to variously reveal or conceal these aspects of my identity in order to build a rapport with my respondents involved certain ethical decisions and implications.

At the time of my field research I was a single woman in my late thirties. Although I was very rarely asked my age (most people assumed I was in my late twenties or early thirties), within a few minutes of our first interaction women would invariably ask me if I was married. When I told them that I was not, they looked with curiosity at my toe rings (usually worn by married women). My explanation that I simply liked to wear them did not satisfy everyone. A few people cited it as an example of the strange things that 'city women' do. For them it was difficult to accept the notion that a woman can do something just because she likes it, even though it is not in accordance with, or is contrary to, cultural practice.

In the region of my fieldwork, girls are usually married off at the age of 18 (the minimum age by law) or earlier. By the time they turn 20 most are mother to at least one child. My respondents found it difficult to believe that I had made a choice not to get married and have children; a woman's life without a family seemed empty in their view (see the chapters by Lunn and Moscuzza and by Staddon in this volume for similar experiences). Since I was going through a turbulent phase in my personal life I initially found such comments emotionally painful, but I soon got used to it.

Though my marital and family status was outside their cultural norms, I think it had a positive impact on our relationship. They felt sorry for me being in my late thirties, single, and without children, and this dissolved the other barriers that existed between us based on factors such as education, class, and caste.

Men with whom I interacted did not openly ask me about my marital status, as it would have been inappropriate to probe a woman about such issues. There was one exception: the brother of one of my female respondents with whom I became good friends. He was about my age and worked in a small factory a few miles from the village. One day he bombarded me with questions such as why I was still studying, how it was to live in a foreign country all by myself, why I was not married, how old I was, and so on. I was not really prepared for such an onslaught of questions. Aware of how people usually look down upon middle-aged, single women, I was tempted not to tell my real age but in this case I decided to be open and honest (though only partially, as I elaborate next). When I told him my age and said that I quite enjoyed my freedom and had no problems being single, he was initially surprised by my answer but then accepted it as one of the indulgences of well-educated, urban, middle-class, upper-caste women living in a Western country. His non-literate sister, who had left her husband, was very pleased at my answer and showed full agreement. This conversation was useful to forge a close bond with this sister-brother duo who later gave me valuable information about land deals in the village.

In rural areas (and to a certain extent in urban areas as well), being a divorced person has a certain stigma attached. It is often assumed that the woman is at fault for the failed relationship and must have done something to offend the husband. In light of this I was very reluctant to talk about my previous marriage and subsequent divorce. Furthermore, the fact that I have not only a stepmother (not so unusual) but also a stepfather (still rare and not a widely socially acceptable occurrence) further complicated my position. However, as part of my study I had to ask probing questions to key respondents about their relationships with their family members and neighbors, and most of the time they confided in me and shared even intimate details. With more closeness to some of these women, I started to feel uncomfortable that I was deliberately withholding details about my own background. My dilemma was whether to be honest or just bear with that nagging feeling. After brooding over it for a month or so, I decided to consult one of my close village friends of many years who had left her husband and was just a few years older than me. She strongly advised me against the honest approach: 'As it is you are a bit of a character and you know that village people think of educated single city women as being loose, why take chances and spoil your impression? It was your past so don't mention it at all.' Worried that it might affect my relationship with my respondents, I decided to follow her advice. I still have not made peace with this choice and I hope to be able to talk about it in the future with some of the respondents with whom I became close.

Another significant aspect of my identity was my educational background. When interviewing non-literate women I emphasized the fact that being educated does not necessarily mean being worldly-wise and that their knowledge of everyday things was valuable too. They often used to make fun of me, saying: 'At this age you are still a student! What's the point of all this education if you cannot find a decent husband and set up a family?'

Income was another contentious issue. I was sometimes asked how much salary I earned or how much money I received for doing the research. If I converted the amount from pounds sterling to Indian rupees I would have appeared quite wealthy and this might have misled people. The fact that I drove to the villages in my car further added to the perception of affluence. I often ended up explaining that I was just a student and had a scholarship to pay my fees and expenses. Usually the respondents who asked about this were aware of the system of educational scholarships and found my reply acceptable. However, I could sense that not everyone was satisfied with the explanation, particularly men who were working in companies and had more exposure to city life, but I tactfully avoided any further discussions on the matter and they were too well-mannered to pursue the discussion. I was also concerned that if the respondents thought that I was well-off, I might receive requests for financial assistance to which I would not be able to respond.

One cultural custom in my fieldwork area was to invite someone to your house after you have visited them in their home. Though I was aware of it, I was reluctant to invite any respondents back to my parental home. My parents live in a modern apartment in a housing complex which includes a small swimming pool and gym; it is also located in an affluent part of the city. However, one of my respondents did happen to visit my home, after which her attitude toward me changed a little bit, albeit for the better. She started mentioning to other villagers how well I managed in their poor living conditions (mainly with reference to lack of toilets and proper beds) in spite of being used to a very comfortable lifestyle. This raised my profile in the village and facilitated an enhanced interaction with my respondents.

I also found that I could intentionally play the well-off urban dweller's card in certain circumstances to create a favorable impression and obtain valuable information, particularly from some government officials. It also made some of my respondents feel important to interact with me. However, I did realize later that resorting to my class position might have helped to perpetuate the existing power relations.

Although there were many aspects of my identity that I could choose to reveal or conceal for the purposes of developing rapport with respondents, there was one aspect that was less flexible. Being a woman in a traditional rural society meant abiding by certain social norms, and I was aware of the 'appropriate' behavior for me as a female. Following those norms meant that I could not hang out with men at tea stalls sharing cigarettes to gather useful information about local politics, land deals, and so on. However, I did manage to make up for that by meeting them at more respectable places such as the local revenue office.

Though the men in my sample villages were all respectful to me and well-behaved I found toward the end of my stay that the local male leaders had instructed my key female respondents to accompany me at all times. Though ties with the village are beginning to loosen up, villagers are still careful to safeguard the reputation of their villages. Also the fact that I was from an urban, middle-class, higher-caste background might have persuaded many to keep the 'respectable' distance.

Madison (2012: 8) emphasizes that 'positionality is vital because it forces us to acknowledge our own power, privilege and biases just as we are denouncing the power structures that surround our subjects.' Certainly my experiences made me reflect in more detail on aspects of my identity. I saw how some aspects of my identity were immediately obvious while others could be divulged as desired; some parts of my identity were less important in certain circumstances yet came to the fore at other times; and some characteristics of my identity provided a way to break the ice while others were a barrier. Though I had wished to be transparent and open while sharing personal information I could not always do so, fearing its undesirable effect on my 'image.' Embedded within this were a number of ethical dilemmas that other researchers may like to consider in terms of whether it is right to conceal or manipulate different aspects of identity for the purposes of accessing respondents.

To probe or not to probe?

Strategically revealing or concealing aspects of my identity enabled me to build rapport with my respondents. My identity and positionality was also significant when extracting research data from my respondents. Ownership of land has become a contentious issue, particularly in areas where land values have risen sharply. Within a family, opinion about selling ancestral land is often divided with one member wanting to sell it off and another obstinate about keeping it. Thus my research topic entered into personal, sensitive, and potentially controversial areas (see also the chapters by Brooks, Day, Skinner, and Taylor in this volume) and I had to strive to ensure that my approach was ethical when extracting information on these subjects in order to minimize any potential upset or harm that my questioning may cause.

I was sometimes asked why I was interested in knowing about these issues, but my replies about its relevance to my research seemed to satisfy them. My familiarity with the field area and network of contacts definitely helped me to build a sense of trustworthiness among my respondents. Also the fact that I was usually accompanied by a local woman may have put them at ease. Furthermore, some of the respondents found it easier to voice their grievances since I was an outsider to their community.

One of the strands of my research was to ask women if they were aware of their legal inheritance rights in land ownership and whether they claimed them. I came across a few cases where there was a dispute between sister and brother over the share in their deceased father's land. It was important for me to understand the dynamics between siblings (against the backdrop of a patriarchal value system) as a means of assessing more acutely the reasons which made women claim their legal rights. Discussions with a range of respondents revealed that in many cases it is the husband who puts pressure on his wife to claim her right in parental property, often against her wishes. I was aware of the fondness with which women usually talk about their parental home, so before asking sensitive questions about inheritance rights I often asked them about their life before marriage, their parental village, if they found it difficult to adjust to the new place after marriage, and so on. The fact that I was also a woman made them feel comfortable. As one of my respondents commented: 'We do not mind talking to you freely because after all you are also a woman – one of us!'

In a few cases answering questions about disputes related to parental property proved to be a painful experience for the women, as the scars were still raw and even the mention of their brother's name made their eyes swell with tears. For my study, such respondents were important as they could have provided me with useful insights; but with those respondents who found it emotionally difficult, I chose not to explore it further as I felt that the respondent's emotional wellbeing took priority (see also Day in this volume). In such cases I made an attempt to collect the information from other family members and friends.

I also came to realize that my questioning of women could be harmful. In one village I was cross-checking data about land ownership with a woman and a male villager was listening intently to our conversation. When she answered my question about the land owned by her family, he immediately started questioning her about her reply. In fact, there was a court case going on between her husband and his cousin about the ownership of a certain plot of land. The man suspected that she was trying to mislead me by including that piece of land under her family ownership. The exchange, which initially sounded like friendly banter, gradually became a grilling session resulting in the exit of my female respondent looking stressed. It is considered rude for a young married woman to argue with her male relatives, especially those from her husband's side. The fact that she was in her early thirties and married to a man who was distantly related to the interrogator made her quite vulnerable. I did not get a chance to find out if she faced any problems as a result of my questions, but it made me realize that I needed to be more sensitive about the roles and expectations of women in rural kinship networks. The incident left me feeling responsible for putting my respondent in a difficult spot.

Since asking women probing questions related to land ownership could have put them in a tricky situation as described above, I decided that it would be better to address such questions to male members of the family. In any case, men are expected to have more information on land-related matters so it was seen as obvious that I should direct any such questions to them. Had a local woman asked similar questions she would have been considered arrogant, but the fact that I was an educated urban woman might have helped in eliciting the information.

Some women asked me if my family owned any land and wanted to know if I had inherited any share in my family property or whether my share had gone to my male cousins (as I do not have a brother). A few male respondents also asked me if I thought it was fair that married sisters can also claim their share in parental property.[8] These questions certainly caused me to question things in my own family life with regards to land ownership that I had previously taken for granted. For example, discussions about the inheritance rights of women made me think about how unusual it was for my maternal grandfather to favor his married daughter over his son in his will, and also how my father's sisters did not get any share in the ancestral family property since they were married. While divulging your own views may help to build rapport, as discussed in the previous section, you also need to consider whether it might shape people's responses to your questions.

While a researcher's primary goal is to collect relevant data, it is vitally important to be sensitive to the consequences of the enquiry for respondents' safety, privacy, and emotions. Other researchers studying subjects that are sensitive, personal, or controversial should carefully consider how to safeguard vulnerable respondents, which may involve reworking the approach or method of data collection.

A personal journey

> Doing fieldwork is a personal experience. Our intuitions, senses and emo-
> tions are powerfully woven into and inseparable from the process.
>
> (Madison 2012: 9)

Although my doctoral research examined the issues faced by rural village
women, the fieldwork experience was a very personal journey for me to find
solace, to overcome personal losses as I made new friends, to challenge and
examine my biases and assumptions, and to understand how I too am shaped by
socio-cultural factors. Reflecting on the influence of my identity on those around
me and on the research process was also a significant part of this personal
journey.

In this chapter I have shared my experiences of the shift from an objective
approach to research to a subjective and reflexive one. Though I had previous
experience of working in rural areas, my doctoral fieldwork gave me an opportunity to understand and reflect more on my positionality and the nuances of
power relations. Although the fieldwork was conducted at 'home' in a location,
culture, and language that I was familiar with, I discovered the complexity yet
significance of identity. My gender, age, caste, class, marital status, and educational background all shaped my interaction with respondents and I had to constantly negotiate aspects of my identity according to the situation. As a
researcher, I believe in being open and transparent in principle. However, it was
disconcerting to have to accept that in practice I manipulated information about
my own identity and family background when it was required in order to build
rapport and gather information; this involved strategically concealing or highlighting certain aspects of my identity. This resonates with the chapters by Dam
and Lunn and by Kovacs and Bose in this volume, who all deliberately manipulated aspects of their identity during fieldwork to access respondents and elicit
data.

As Madison (2012: 100) says:

> To examine one's own life and intentions – to question and observe the self
> – in the process of questioning and interacting with others is an ethical
> stance because it requires consistent self-evaluation and monitoring relative
> to our integrity, effectiveness and political commitment toward the end of
> helping make life more worth living.

To observe rural villagers' everyday survival struggles from close quarters
was unnerving, making me realize how privileged I am. The fact that I could do
very little to improve the situation for my respondents sometimes made me frustrated. However, I hope that the self-reflection and learning that my fieldwork
prompted will, in the longer term, facilitate in working toward the betterment of
the situation for the people whom I studied.

Recommended reading

Chacko, E. (2004) 'Positionality and praxis: Fieldwork experiences in rural India,' *Singapore Journal of Tropical Geography*, 25 (1): 51–63.

This paper shows how the multiple subject positions and identities of both scholar and subjects affect access to informants and the tenor and outcomes of encounters.

Huong, N. (2007) 'Anthropology at "home" through the lens of intersubjectivity,' *Medische Anthropologie*, 19 (1): 23–38.

Based on the premise that the degree of subjective experience is an inherent part of conducting research, this paper illustrates the nature of doing fieldwork within one's own culture.

Sultana, F. (2007) 'Reflexivity, positionality and participatory ethics: Negotiating fieldwork dilemmas in international research,' *ACME: An International E-Journal for Critical Geographies*, 6 (3): 374–385.

Drawing on fieldwork in Bangladesh, this paper discusses how concerns over reflexivity, positionality, and power relations are even more important in the context of multiple axes of difference and inequalities.

Notes

1 In many parts of India, surname usually indicates a person's caste.
2 *Dalit* is a term used for those who were formerly referred to as 'Untouchable'; it refers to the most oppressed and exploited sections of the caste system (Omvedt, 2011).
3 Brahmin is a priestly caste which traditionally falls at the top of the caste hierarchy in many parts of India.
4 There is a lot of discussion in recent anthropological literature on the extent to which caste forms the 'foundational' element in Indian society.
5 In fact, in recent years there has been revived enthusiasm among urban Brahmins for organizing caste-based global conventions, and my hometown of Pune is the head-quarters for this.
6 See Osella (2008) for the significance of the social and cultural meanings of food in shaping relationships in the field.
7 Rural people often use the word *baman* for Brahmin; in recent years it has acquired a derogatory sense.
8 As per a recent amendment to the Hindu Succession Act 1956, married daughters can claim their share in ancestral property.

References

Chacko, E. (2004) 'Positionality and praxis: Fieldwork experiences in rural India,' *Singapore Journal of Tropical Geography*, 25 (1): 51–63.
Gilbert, M. R. (1994) 'The politics of location: Doing feminist research at "home",' *The Professional Geographer*, 46 (1): 90–96.
Hammersley, M. and Traianou, A. (2012) *Ethics in qualitative research: Controversies and contexts*, London: Sage.

Madison, D. (2012) *Critical ethnography: Method, ethics and performance*, Los Angeles: Sage.

Maynard, M. (1994) 'Methods, practice and epistemology: The debate about feminism and research,' in M. Maynard and J. Purvis (eds), *Researching women's lives from a feminist perspective*, London: Taylor and Francis.

Omvedt, G. (2011) *Understanding caste: From Buddha to Ambedkar and beyond*, New Delhi: Orient Blackswan.

Osella, C. (2008) 'Introduction,' *South Asia: Journal of South Asian Studies*, 31 (1): 1–9.

Srinivas, M. N. (1997) 'Practicing social anthropology in India', *Annual Review of Anthropology*, 26: 1–24.

Srinivas, M. N., Shah, A. M. and Ramaswamy, E. A. (eds) (2006) *The fieldworker and the field: Problems and challenges in sociological investigation*, 2nd edn, New Delhi: Oxford University Press.

Sultana, F. (2007) 'Reflexivity, positionality and participatory ethics: Negotiating field-work dilemmas in international research,' *ACME: An International E-Journal for Critical Geographies*, 6 (3): 374–385.

9 First impressions count: the ethics of choosing to be a 'native' or a 'foreign' researcher

Two tales from fieldwork in India

Rinita Dam and Jenny Lunn

> 'Sir, that researcher has come back to ask you some questions.'
> 'Ah, tell her I'm busy and to come back tomorrow.'
> 'Ah, show her in and bring some *chai* [tea] straight away.'

These were typical exchanges during our fieldwork in India, with Rinita tending to hear the first response and Jenny the second. We feel that people's first impressions of us, based on certain aspects of our identities – particularly skin color,[1] language, and dress – and the assumptions that they made on the basis of this, were highly significant in influencing our research, particularly in terms of access to research participants and data.

Every researcher has a range of personal and social identities which comprise who they are, ranging from gender, age, and nationality to class, qualifications, and religion. In the Indian context there is the added dimension of caste which, for 'natives,' is usually distinguished by surname (see also Godbole in this volume). During fieldwork, certain aspects of a researcher's identity are more significant while others are less important, influenced by a combination of factors including the research topic, location, methodology, and relationship with respondents. Chacko (2004: 53) reflects that 'permutations of these … groupings produce multiple identities which may mesh well or tangle awkwardly in any given setting.'

The 'ideal' for social science research is an objective approach where the researcher is detached from the research subjects. However, the reality is that researchers are human beings and their research involves interacting with other human beings. Thus a researcher's identity and the way in which this is received and perceived by others cannot fail to play a role in influencing the research in a variety of ways, including developing rapport and building trust, gaining access to respondents, and gathering data.

Researcher identities can be divided into two types variously labelled in the literature; for example, 'prescribed' and 'ascribed' (Herod 1999) or 'real' and 'perceived' (Chacko 2004). The former are aspects of identity that are visibly undisputable, such as skin color and gender; the latter are aspects of identity that people assume based on appearance or behavior, such as clothing or jewellery.

Thus there are some aspects of identity that are within our control and others that are beyond our control.

From this, various ethical dilemmas arise. In terms of those aspects of our identity that are within our control, is it ethical to deliberately manipulate – either by revealing, concealing, emphasizing, or down-playing – certain attributes for the benefit of the research? In terms of those aspects of our identity that are out of our control, what is the ethical response when the identity that people ascribe or assume of us is incorrect yet has a significant influence (positive or negative) on the research?

We carried out doctoral fieldwork in the Indian city of Kolkata at the same time. Our respective supervisors, who were long-standing acquaintances, had discovered that they each had a student heading to Kolkata and put us in touch. We were able to meet several times during our fieldwork and, during these conversations, discovered that we were having two very differing experiences, particularly regarding access to respondents. We came to suspect that this was being influenced by our identity and the positionality it created. We were both female, young(ish), British doctoral students from UK universities, so those aspects of our identity (i.e., gender, age, nationality, and professional status) were comparable. Instead we felt that ethnicity – within which we incorporate skin color, language, and dress – were much more influential. This chapter describes our contrasting experiences and explores the ethical dilemmas that we faced as a result.

Full of good intentions

Rinita

My research focused on the impact of HIV/AIDS on the lives of the poor. Specifically, it explored the coping strategies of individuals and households living with HIV/AIDS, as well as their access to treatment and support services offered by different types of healthcare providers. I chose the city of Kolkata for my fieldwork for three main reasons. First, I had previously worked for a community-based organization in the city and through this was aware of how some not-for-profit organizations operated, had some contacts within the sector, and also had an understanding of the issues faced by poor people living with HIV/AIDS. Second, my mother tongue is Bengali, the local language, which I assumed would be advantageous, enabling me to carry out in-depth interviews without the need for an interpreter. Third, I had a personal connection; it is the city of my mother's birth and I spent a short period of my childhood there.

For these reasons, I chose to locate my field research in Kolkata and did not consider other possibilities. I assumed that, by doing field research in a place I already knew and where I had some contacts, I would be at an advantage. However, I would urge other researchers to question such assumptions – as my own research experiences relayed in this chapter will demonstrate (see also Gent in this volume on making assumptions about the relative 'easiness' of research in a place where you already speak the language).

My initial plan was to select a low-income neighborhood, to identify available healthcare and other relevant services, and then to assess the overall health and social resource package available to people living with HIV/AIDS in that neighborhood. However, during my pilot study it became clear that it would not be possible to identify people living with HIV/AIDS even with good knowledge of the neighborhood. Asking people directly about their HIV status, or that of household members, was simply unacceptable due to widespread stigma and the lack of privacy within households. Even if I knew which individuals were living with HIV/AIDS, it would have been unethical research practice to spot-interview them. Thus I realized that the only way to identify potential research participants would be through local organizations working in this field.

I started by using internet searches to identify non-governmental and governmental organizations providing care, support, and treatment to people living with HIV/AIDS within the city of Kolkata. I hoped to be able to work with these organizations in two ways: first, to interview senior staff in order to understand the delivery of services to people living with HIV/AIDS; and second, to ascertain whether they would be willing and able to help me identify people living with HIV/AIDS with whom I would carry out in-depth qualitative interviews for the final phase of my research. In the end, I worked with four different organizations and interviewed 59 people living with HIV/AIDS.

Jenny

My research was about religious organizations engaged in social development activities, examining which aspects of their beliefs and scriptures motivate their engagement with poor and underprivileged members of society, and the extent to which religion is or is not integrated into their outreach activities. As for Rinita, Kolkata was the uncontested choice of fieldwork location for me. My doctoral research had grown out of almost a decade of interaction with the city, during which time I had volunteered for a range of religious and non-religious organizations and had carried out research for my master's degree looking specifically at the social development activities of Christian organizations. My doctoral research was an extension of this, widening out to ten different religions. Like Rinita, I was familiar with the city after repeated visits and had a working knowledge of the particular NGO sector and contacts therein. Although I did not have much vernacular language capacity beyond basic words and phrases, I assumed that research would be 'easier' in a place I already knew and that I would not 'waste' too much of my allocated fieldwork period in getting to know the place and establishing contacts. However, things never go quite as you imagine.

My plan was to examine the development activities of religious organizations belonging to ten different religions. I expected that the government or some other body would have an easily-accessible list of all registered NGOs in the city that I would use as a pool from which to select a sample. However, I could not find such a publicly-available list so I spent the first few months of my fieldwork period searching the internet and creating my own as-comprehensive-as-possible list of NGOs in the city, and from that ascertaining what proportion had a religious orientation. From that list, I then planned to select 50 religious organizations representing ten different religions and to conduct semi-structured interviews with the most senior figure in order to explore organizational motivation, structure, and practice. In the third stage, I planned to select just three of the 50 organizations for in-depth study through additional interviews and participant observation.

Reality bites

Rinita

While Jenny and I had a similar methodological approach, our experiences at the outset of our fieldwork were quite different. Prior to my pilot study, I had contacted relevant people in NGOs and government by email and telephone to request a meeting once I was in Kolkata. However, I received no replies. Once I arrived in India, I called these organizations again to request face-to-face meetings. On some occasions I managed to arrange a meeting, but once there I would be told that they were now unavailable to talk to me either due to prior commitments or because they had forgotten they had scheduled a meeting with me. I would then be told to 'come back tomorrow' (also see chapters by Chopra and Perera-Mubarak in this volume for similar frustrations). Sometimes I tried an alternative strategy, which was to arrive at offices without a prior appointment. This was also met with rejection, as I would be told that nobody was available to talk to me and was advised to make an appointment with the relevant person before turning up at the office. I felt as if I was going around in circles!

However, my research relied on gaining access to these organizations, both to learn about healthcare provision and to identify people living with HIV/AIDS. I simply had not anticipated that I would struggle to gain access to senior staff working within NGOs, public hospitals, and other government agencies. Furthermore, once I did gain an 'audience,' people were not particularly helpful with my enquiries. I felt unproductive and disheartened and, on many occasions, questioned the very reason for doing the research. It was so bad that about halfway through my pilot study, I nearly gave up.

Jenny

I started by using my existing contacts in organizations that I had interacted with in the past. I then intended to use a snowballing technique by asking interviewees to recommend people from other organizations for me to interview. I also thought this best for the Indian context, where personal introductions are important. However, this was based on my assumption that the NGO world – and specifically religious NGOs – would be well networked. This proved to be true to an extent within religious denominations, but much less so between denominations and religions. Thus in order to access organizations of particular religions or denominations where I had no existing contacts or personal recommendations, I had to do 'cold calling' either by email or telephone.

I was stunned by the response to these cold calls. On the phone I was put straight through to the most senior person after only a brief introduction of myself and my research. Numerous times the invitation came to 'come over to my office right now.' I had emails answered within hours inviting me to meet the following day. People even dispatched their drivers to collect me and take me to interviews. The longest delay between initial contact and interview was a couple of days, and that was only because of my prior commitments.

I was amazed that the heads of organizations (e.g., general directors, secretaries, presidents, trustees) could take time out of their busy schedule to speak to me. I was immediately invited into the 'inner sanctum' of their offices while dozens of people waited in line outside for an 'audience.' Distractions were waved aside as I interviewed them. I found people generous with their time and their information. The offers came thick and fast to complement the interview with visits to projects and beneficiary communities. One could speculate that some religious motivation of doing good to others or being hospitable was the driving force behind the positive responses but, as the next section will show, we suggest that this was not the most significant factor.

Learning the hard way

Rinita

Having repeatedly received the cold shoulder from both governmental officials and NGOs, I began to suspect that there was something about me or my research that they did not approve of or like. When I was eventually granted some interviews, I asked the reason for their negative responses to my previous requests. I was so fed up by this stage that I did not feel in any way bashful or embarrassed in questioning them about this.

I received the response that I was perceived to be 'one of their own.' First contact over the telephone or by email had revealed my Indian name (and with my surname, an indication of my caste background). Then first contact in person had affirmed my 'native' identity. I was told that because I did not *look* like a 'foreigner' and had spoken to them in Bengali, they treated me in the same way they would treat any other person born and bred in Kolkata. Cultural sensitivity is one aspect of good ethical behavior and I thought that I had done the 'right' thing by speaking to people in their native language and wearing clothing appropriate to the setting (i.e., *salwaar kameez*), but it seemed that my efforts to appear 'native' were part of the problem.

When I presented my British passport to prove my 'foreigner' status, people became very curious about my background, questioning me about my parents' professions, where I had been born, how I was able to converse in Bengali, and at what age and why I had left India. It was almost as though they were interrogating me to determine how genuine my foreignness was.

I also probed these people further about what they meant by me not *looking* like a foreigner, to which they responded that I was not a *real* foreigner, i.e., a Caucasian. When questioned about why they would treat foreigners differently, their response was that they did not want to appear rude or unwelcoming; they were concerned about making a bad impression and fearful of negative press. This was often a prelude to a long speech about their fascination for all things foreign and nostalgia for the British Raj.

I was glad to have an explanation for the difficulties that I had experienced in gaining an audience with these NGO managers and government officials. I was not particularly concerned about what they actually thought of me as a person, but wanted to use this knowledge to improve my access to others, particularly because this was crucial for the next stage of my research.

Thus I returned to Kolkata for the second period of my fieldwork with businesslike, Western-style clothes. When meeting senior people from government agencies and NGOs, I started the encounter by whipping out my British passport. I spoke only in English and pretended not to know a single word of Bengali. I was determined to make a point about how foreign I was. I repeated these actions in a ritualistic fashion every time I had to interact with senior people. I could not change the color of my skin, but the overt emphasis on my nationality and the change in language and dress, made the second phase of my fieldwork much more productive.

However, the second part of my data collection – interviewing people living with HIV/AIDS – was quite a different story, in contrast to Jenny's experience as related below. Having been granted access to respondents, I once more reverted my identity: I spoke in Bengali and dressed in Indian clothes. All of the people I interviewed were extremely generous with their time, with some interviews lasting over an hour. I also found my participants very receptive and open to my questions despite their personal and sensitive nature.

Once the interviews were over, some participants had time to chat and I was often asked about my background and about what my parents did for a living. As with the NGO managers and government officials, this questioning was a mechanism to 'place' me on the social and class spectrum, but these people also seemed to ask out of interest. Many were bemused by my ability to speak Bengali – albeit with an English accent – and curious about why I chose to return to India to conduct a study about poor people living with HIV/AIDS. After all the difficulties I had encountered, I very much appreciated spending time with people who were so accepting of me and of my research.

Jenny

Following my very successful 50 interviews with leaders of religious organizations, I selected three organizations for more in-depth study. I hoped to complement the 'official' views of senior organizational representatives with the view from beneficiaries. However, in contrast to Rinita, here my lack of capability in the local language was my weakness. Group interviews with beneficiaries were set up by field officers of the three organizations who were also present to translate. Herein lay two problems. First, as with any research which utilizes interpreters, there is the difficulty of knowing whether the words and concepts that you speak and the respondents' replies are literally translated or 'mediated' by the interpreter (see chapters by Gent and Leck in this volume for more detailed discussions about research in a second language, working with interpreters, and translation). Second, the presence of the organizational representative may well have influenced the responses from beneficiaries, who may not have wanted to criticize the work of the organization or the services that they received. Thus, I could use some of the factual data from these group interviews but could not rely on people's opinions and views as being genuinely independent and representative voices. Therefore this stage of my research, which was originally intended as the centerpiece, eventually took lesser significance in my final thesis.

Unexpected outcomes

Rinita

The reluctance of many organizational officials to provide assistance meant that I was not able to undertake a systematic review nor to conduct an evaluation of service providers. Furthermore, my lack of access to official data meant that I could not generalize findings across the city. As a result of officials' initial assumptions about my 'nativeness,' the focus of my research inadvertently shifted. In some ways I was disappointed with the eventual emphasis of my research, as I had wanted it to have more policy relevance. I had not originally intended to conduct predominantly qualitative fieldwork among the poor; yet my experience of interacting with people living with HIV/AIDS who, unlike officials, had not pre-judged me on my appearance was a humbling and rewarding experience and a definite highlight of my fieldwork.

Jenny

The extreme helpfulness of organizational officials generated a wealth of interview transcripts and written data far beyond my expectations, while the responses from beneficiary interviews brokered through interpreters resulted in much less useful data. My 'foreignness' was influential in both gaining access to elites and having limited access to beneficiaries. As for Rinita, this shifted the original focus of my research, resulting in more of an emphasis on organizational structure and performance than on implementation and impacts at the community level. Overall, I do not feel too disappointed with this outcome, since the privileged access to religious elites afforded by my 'foreignness' was definitely the highlight of my fieldwork.

Exploring the ethical dilemmas of manipulating identity and challenging assumptions

> Given the strict security and time constraints that most elites operate under, researchers often find themselves with only a brief window of opportunity to convince those from whom they seek information that such an endeavour is worthwhile. In this brief encounter, a researcher's positionality whether perceived or represented often has a crucial impact upon whether he/she is granted an interview.
>
> (Mullings 1999: 339)

Although our research topics were different, we shared a fieldwork location and there were many parallels in our respective methodologies, which allow us to compare our research experiences. We both began our research with interviewing elites and the above quote by Mullings rang true with us in terms of the significance of first impressions. In those 'brief encounters' people made judgments and assumptions about our identity that shaped not just one interview experience, but our wider fieldwork experience and the direction of our research as a whole. Specifically, we feel that our perceived 'native' and 'foreigner' identities were of foremost importance. This is not to say that skin color, language, and dress were the only aspects of identity that influenced our experiences – age and gender were certainly at play for both of us, while religious affiliation was also a factor in Jenny's research. Our respective research topics may have also had an influence on positionality and access; for example, we could speculate on differences if Rinita had been researching a health issue with less sensitivity or if Jenny had been investigating a more controversial issue, such as religious and communal violence.

Our reflections focus on the ethical dilemmas arising from our 'native' and 'foreigner' identities and our respective reactions to this ascribed identity. The first dilemma concerns whether it is ethical or unethical to manipulate selected aspects of our identity in order to increase the success of our research. The second dilemma revolves around the ethical response to incorrect assumptions about our identity when these are either damaging or beneficial for the research. Of course these ethical dilemmas are applicable to any aspect of researcher identity and some other cases are discussed elsewhere in this volume (e.g., the chapters by Godbole and by Kovàcs and Bose).

In our case, Rinita was visibly (skin color) a 'native' and deliberately chose to adopt local cultural behavior (language and dress). Because of these factors, people made the assumption that she was 'one of them' and based their responses to her enquiries on this. Having discovered that her perceived identity was negatively affecting her research plans, Rinita deliberately changed those aspects of her identity that she was able to alter (language and dress) and openly challenged people's assumptions that she was Indian by showing her British passport. It had a striking impact on people's responsiveness and helpfulness. She had initially tried to behave in an ethical manner by adopting local customs as a means of showing respect; thus can her decision to deliberately change her identity be regarded as 'unethical'?

The literature on the insider/outsider dichotomy is probably the most instructive for exploring Rinita's dilemma. Other researchers have faced a similar situation of having a visible identity that positioned them as 'natives' or insiders in the research location while other aspects of their identity distinguished them as 'foreigners' or outsiders (e.g., Mullings 1999; Mohammad 2001; Tembo 2003; Chacko 2004).

The binary position of insider or outsider has been widely criticized as simply not a reflection of reality. During fieldwork, Mullings (1999) was an insider in one situation yet an outsider in another and describes unstable boundaries and dynamic positionalities. Similarly, Mohammad (2001) describes her identity and positionality as highly ambiguous, constantly shifting and negotiated. These researchers and many others have used this flexibility of position for the benefit of their research including to develop rapport with people, to gain access to communities, and to gather data. However, there is little reflection in the literature on whether it is ethical or unethical to capitalize upon ambiguity of position and flexibility of identity.

Uncertainty surrounds whether deliberately manipulating your identity constitutes deceit. While some ethnographic research is covert and depends on deception (and tends to be subject to more stringent ethical procedures), most development researchers are not acting under cover. Ethical guidelines call for transparency in our dealings with research participants. One aspect of this is informed consent, which includes a full explanation of the purpose of our research, being clear about people's involvement, and assurances of confidentiality. Thus does it follow that being transparent about our research procedure also means being honest and open as people? (Again see Godbole in this volume, who faced similar dilemmas about how much of her personal identity to reveal or conceal.)

There is no easy solution to this dilemma, but we suggest that deliberate manipulation of identity can be acceptable if you are fully aware of it and if it does not cause harm to you or your participants. Rinita faced a situation where her attempts to behave in an ethical manner jeopardized the success of her research project, and changing aspects of her identity to shift from a 'native' to a 'foreigner' positionality was her only remaining means of ensuring access to participants and the continuation of her fieldwork. Pretending not to know the local language was a small act of deception, but neither that nor changing her dress and demonstrating her British nationality caused any harm.

Jenny's experience was a complete contrast to Rinita's but was not without ethical dilemmas. She was visibly (skin color) and behaviorally (language and dress) a 'foreigner.' Nor did she resist any assumptions that people made on the basis of her appearance. This was a deliberate tactic based on previous experience, knowing that in India there is still a certain regard for all things British. Her preferential treatment and fast-track access to elites for interviews was largely a result of this visible and 'performed' foreignness.

As Rinita did, Jenny could have followed ethical guidelines for culturally-sensitive research by learning the vernacular language and by dressing in the customary manner. Was deliberately not doing this therefore unethical? However, the heart of this ethical dilemma is the fact that the aspects of identity that Jenny exploited to ensure the success of her research were all related to her 'foreignness.'

Some of the literature on postcolonialism is worth reflecting on to explore this ethical dilemma. There has been a long-standing debate on the appropriateness of 'First World' academics doing research in the 'Third World' (e.g., Sidaway 1992; Potter 1993; Madge 1993; Sidaway 1993; Porter 1995) and whether it constitutes a form of neoimperialism. The postcolonial critique of development caused researchers to become more reflexive about their positionality (e.g., Kapoor 2004) and to search for different intellectual frameworks and methodological approaches for conducting research in the Global South in order to transcend postcolonialism (e.g., McKinnon 2006; Raghuram and Madge 2006).

Many twenty-first century researchers will feel temporally distant from the colonial period; they may also disagree intellectually and politically with neocolonial discourse and practice. However, the reality when conducting fieldwork in many countries of the Global South is that colonial histories and politics can still influence contemporary research (Skelton 2001; McKinnon 2006). For example, Besio (2003) conducted research on the role of women and children in the mountain portering economy in Northern Pakistan. She assumed that gender would be the most important aspect of her identity for carrying out the fieldwork. However, the local community and research participants saw her primary identity as colonial/postcolonial. Furthermore, she could not resist this identity that was imposed on her, partly because nothing would have changed their views and partly because the very act of resisting or challenging it would have been reinforcing those very colonial identities of power.

In a similar vein, Jenny's identity as white and British was not of any direct relevance to her research topic yet, on first contact, this visible identity, which was verified by language and dress code, created an unspoken positionality and shaped people's subsequent reactions and interactions. She did not try to downplay this aspect of her identity because it proved useful for the research in terms of gaining access to elite members of society, entering circles of trust, and extracting information. This capitalization on her postcolonial positionality could be viewed as perpetuating the privileged and exploitative relationship so characteristic of colonial relationships. However, as Besio (2003) points out, the dilemma is that confronting this positionality might have acted to further reinforce it.

Perhaps the only way to leave this complicated debate with a conscience intact is, as Rinita did, to consider whether this behavior caused any harm to the researcher or the participants. In all likelihood it did not cause any harm, as it was the participants who ascribed the postcolonial identity upon Jenny and decided that it was significant. However, there may be harm caused to others as experienced first-hand by Rinita, who saw how 'foreign' researchers are privileged over 'native' researchers.

We hope that our contrasting experiences and unresolved ethical dilemmas will allow other researchers to reflect on the following: the choice to do research in a familiar location, which it is typically assumed will put you at an advantage; the potential impact on your research of first impressions based on your visible identity; the fluidity between 'native' and 'foreign' identities; and some of the dilemmas faced in projecting particular identities and dealing with people's assumptions. Both of our experiences leave us with the very difficult question of whether, or to what extent, ethics can be compromised in order to ensure the success of your research.

Recommended reading

Chacko, E. (2004) 'Positionality and Praxis: Fieldwork Experiences in Rural India,' *Singapore Journal of Tropical Geography*, 25 (1): 51–63.

Chacko's reflexive account of fieldwork in India explores the 'uneasy balance' between being both an insider and an outsider. She shared ethnicity, language, and gender with her research participants but found that they came from a completely different life-world and cultural framework. This caused her to reassess areas of her life, her personality, and her intentions, and also to question the appropriateness of her field research.

Mohammad, R. (2001) ' "Insiders" and/or "outsiders": Positionality, theory and practice,' in C. Limb and M. Dwyer (eds), *Qualitative methods for geographers: Issues and debates*, London: Arnold.

Mohammad provides some interesting reflections on four research projects among the Pakistani community in Britain. She discusses the ambiguity and flexibility of her simultaneous insider and outsider positionality and describes how her assumed and ascribed identity and positionality variously affected access, trust, disclosure of information, and so on.

Mullings, B. (1999) 'Insider or outsider, both or neither: Some dilemmas of interviewing in a cross-cultural setting,' *Geoforum*, 30: 337–350.

Mullings describes how her multiple positionalities, particularly gender, race, age, and class, influenced her fieldwork in Jamaica, particularly in terms of data collection and interpretation. She also critiques the insider/outsider binary and suggests that researchers meet research participants in transitory 'positional spaces.'

Tembo, F. (2003) 'Multiple identities and representations: Experiences in the study of people's life-worlds in rural Malawi,' *Singapore Journal of Tropical Geography*, 24 (2): 229–241.

Tembo discusses mobilizing multiple identities to gain access to the research community, reconciling these identities, and using them appropriately in different contexts to gain access.

Twyman, C., Morrison, J. and Sporton, D. (1999) 'The final fifth: Autobiography, reflexivity and interpretation in cross-cultural research,' *Area*, 31 (4): 313–325.

This paper reflecting on field research in Botswana contains various examples of people making assumptions about the researchers' identities based on appearance and behavior. The researchers also describe how their identity shifted through the course of the research and on successive visits, and how this influenced the relationships with research participants and data collection.

Note

1 We have deliberately chosen to refer to skin color rather than 'race' because of the complexity of the latter term.

References

Besio, K. (2003) 'Steppin' in it: Postcoloniality in northern Pakistan,' *Area*, 35 (1): 24–33.

Chacko, E. (2004) 'Positionality and praxis: Fieldwork experiences in rural India,' *Singapore Journal of Tropical Geography*, 25 (1): 51–63.

Herod, A. (1999) 'Reflections on interviewing foreign elites: Praxis, positionality, validity, and the cult of the insider,' *Geoforum*, 30 (4): 313–327.

Kapoor, I. (2004) 'Hyper-self-reflexive development? Spivak on representing the Third World "Other",' *Third World Quarterly*, 25 (4): 627–647.

McKinnon, K. I. (2006) 'An orthodoxy of "the local": Post-colonialism, participation and professionalism in northern Thailand,' *The Geographical Journal*, 172 (1): 22–34.

Madge, C. (1993) 'Boundary disputes: Comments on Sidaway (1992),' *Area*, 25: 294–299.

Mohammad, R. (2001) ' "Insiders" and/or "outsiders": Positionality, theory and practice,' in C. Limb and M. Dwyer (eds), *Qualitative methods for geographers: Issues and debates*, London: Arnold.

Mullings, B. (1999) 'Insider or outsider, both or neither: Some dilemmas of interviewing in a cross-cultural setting,' *Geoforum*, 30: 337–350.

Porter, G. (1995) ' "Third World" research by "First World" geographers: An Africanist perspective,' *Area*, 27 (2): 139–141.

Potter, R. B. (1993) 'Little England and little geography: Reflections on Third World teaching and research,' *Area*, 25 (3): 291–294.

Raghuram, P. and Madge, C. (2006) 'Towards a method for postcolonial development geography? Possibilities and challenges,' *Singapore Journal of Tropical Geography*, 27 (3): 270–288.

Sidaway, J. (1992) 'In other worlds: On the politics of research by "First World" geographers in the "Third World",' *Area*, 24 (4): 403–408.

Sidaway, J. (1993) 'The decolonisation of development geography,' *Area*, 25 (3): 299–300.

Skelton, T. (2001) 'Cross-cultural research: Issues of power, positionality and "race",' in C. Limb and M. Dwyer (eds), *Qualitative methods for geographers: Issues and debates*, London: Arnold.

Tembo, F. (2003) 'Multiple identities and representations: Experiences in the study of people's life-worlds in Rural Malawi,' *Singapore Journal of Tropical Geography*, 24 (2): 229–241.

10 Flirting with boundaries: ethical dilemmas of performing gender and sexuality in the field

Two tales from conservation-related field research in Hungary and India

Eszter Krasznai Kovàcs and Arshiya Bose

No drink, no dinner, no interview.

(Extract from Arshiya's interview notes)

I'm sitting here after three weeks, feeling withdrawn and lost and pretty angry, almost like I don't remember what it's like to stand up for myself anymore. Hel-loooo I am more than my vagina. I feel like I have to laugh at everything, even when I don't understand what's so funny about me being here, asking questions, just because ... what? Because I'm a girl? I don't find their jokes funny. But then why do I keep laughing?

(Extract from Eszter's field notes)

As doctoral students, we travelled to countries that we already knew to varying degrees for our fieldwork. For one of us, 'the field' constituted 'home' (India); for the other, an historical home but with a childhood spent elsewhere (Hungary). The similarities in our experiences in these spatially and culturally disparate 'fields' started a conversation and led to this chapter (in contrast see Dam and Lunn in this volume for differing experiences in India). While India is categorized as part of the Global South, the rapid economic, social, and cultural development seen there is replicated, albeit on a different scale, in eastern Europe which, as a result of its socialist past, still has significant 'catching up' to do with the historically more established members of the Global North. Furthermore, the economic categorizations of Global South and North seem to matter little in our cultural experiences of conducting fieldwork, and hence the inclusion of Hungary in this chapter of a book focusing on the Global South.

Our mutual identification as confident, 'go-getting,' determined women needing to engage and become accepted within our chosen fields and field sites presented the primary challenges; how we individually coped led to the ethical dilemmas and situations recounted here. Herein we argue that one's gender as a researcher *counts*. It counts not only when the field is male-dominated and hierarchical and one needs to gain access and make friends with gatekeepers; but it also counts, albeit differently, at an individual level, when access has been granted and conducting long-awaited interviews can finally take place. These gender dynamics are rarely acknowledged in the 'learn how to interview' classes and texts that most postgraduate students use for training, although an exception and a great compilation is *Taboo: Sex, Identity and Erotic Subjectivity in Anthropological Fieldwork* (Kulick and Willson 1996).

There are, naturally, no hard-and-fast, black-or-white rules for how to handle gender or sexuality in any situation; these are everyday issues and not nominally fieldwork-based. In the following sections we discuss in turn our practical experiences of gaining access. First we explore the gender dynamics in our respective contexts, including the predetermined role of women who were already present, either as co-workers or as locals. Second we recount what happened to us 'in the field.' Third we discuss the ethical dimensions of 'using' or 'leveraging' gender dynamics to one's benefit, for access or acceptance, through the lens of gender performativity.

Explaining the field

Eszter, Hungary

In recent decades, Hungarian society has undergone massive transitions economically, socially, culturally, and environmentally. The role of women in socialist times was one of contradiction, a malaise between the apparently gender-neutral socialist need for all labor (men and women) and a fairly conservative, traditional cultural heritage whereby women married young and became mothers. Although the 'sexual revolutions' of the West in the 1960s did not entirely fail to penetrate the Iron Curtain, it arrived later, restructuring sexual relations and the acceptability of younger people having sex and giving birth to children outside wedlock. Outside of this personal sphere, the communist economic system prided itself in the egalitarian advancement of women, and as a result women were highly represented in management, albeit there was a high prevalence of 'gendered' jobs (Pollert 2003).

In today's Hungary, I would argue that there has been a political, rhetorical, and real 're-masculinization' of traditional gender identities and roles in society over the past few years. This is reflected in the work practices of the state's formal institutions, hiring guidelines and biases, the erosion of state-provided childcare services (c.f. Metcalfe and Afanassieva 2005, Weiner 2009), childcare leave provisions, and a political and social environment that promotes traditional family structures. One extreme example are the words of an MP of the governing Fidesz party spoken in Hungarian Parliament during 2012 who stated that 'alongside emancipation and finding themselves women forget to give birth' and that 'women should start thinking about work once they have raised four or five children.'[1] Such attitudes and approaches to women's role in society are not rare in Hungary today, reflecting in essence that it is governed by a conservative-Christian coalition. In many ways I have become familiar with and encountered similar attitudes through my own experiences of working and living there, some of which I recount here. These are of course personal and specific to my time 'in the field.' I do not in any way mean to suggest that the dynamics, gender roles, or sexism that I describe are pervasive and true of the whole country nor across the conservation and agricultural sector.

My doctoral research examined the nexus of conservation and agricultural approaches to land management. As part of my fieldwork I travelled the length and breadth of Hungary, visited all ten National Park Directorates, and established two 'local level' case study sites, wherein I met a number of farmers and their families. In fact, by this I mean that I met a lot of men. Almost all farmers without exception were male. In most cases where I interviewed women, it was because their husbands were unavailable, or because the women completed the administrative/marketing side of their farming enterprise, and it was thought that they would be better placed to answer my questions. None of these women could answer practical farming questions about particular practices relating to physical work in the fields; it was a case of, 'Oh, my husband does that.' This has two significant consequences for my research. First, as a result of farming society being male-led, male farmers interact with and are used to working and engaging with men similar to them. The presence and roles of women in this are as supportive family members, not as fellow managers with similar work issues and anxieties. Second and pursuant to this, to thus enter such societies as a woman interested in farm management and its issues is, from the first instance, more difficult.

In National Park Inspectorates, my interviewees were roughly even between men and women, but even here the work divisions were highly gendered. Of the coordinators of the conservation program I studied, although a significant proportion were female, they were almost without exception performing administrative and office work. The work of a National Park Ranger, on the other hand, was described by most of these interviewees as an outdoor, field-based position that came with driving a four-wheel drive and carrying a gun. Most women agreed that Rangers should be men as it was 'dangerous' and 'not female work' and women would be less able to command respect if confronted by trouble. These perceptions transcended urban/rural divisions, and in my personal observations of National Park Rangers that are women (they do exist!), there is strong gender performativity at play as many conform, adopt and live roles and attitudes that lead them to be perceived as 'one of the men' (Butler 1990).

I am not one of these women. I am slender, loud, and opinionated and have a big laugh. I would not know where to begin if I needed to masculinize myself; disappearing and 'becoming invisible,' as recommended in many anthropological enquiries I also find difficult. However, I found that this was not necessary in a field composed of men. Being visibly *female* has distinct advantages – but only if you behave and conform to rules and requests determined by men in the 'right' positions. In three different National Parks, I encountered the phenomenon of the male Ranger that was sexist in jest, and because all references to my femininity or vagina were *a joke*, it wasn't being sexist. Here, female 'emancipation' and political correctness as a whole was mocked. In essence, I felt that due to their isolation these men got to make their own rules.[2] Thus women in these specific spheres were categorized: laugh and play along, and you increase the chances of being liked and accepted. I found that stating the unacceptability of jokes caused friction and tensions that were not conducive to gaining access or building trust with gatekeepers. Almost all women that I encountered 'played along,' which was very challenging to me, and apart from a few fellow research colleagues in Hungary, I did not encounter women that would even admit that such positioning of females, conservationists or not, was problematic, let alone sexist.

Arshiya, India

My 'field' was a strange and mystical place like no other I have ever encountered in India. Somewhere between the rolling Western Ghats and the endless sea of coffee plantations lay a deeply complex social fabric. My research explored how coffee growers managed natural ecosystems on their plantations, including use of chemical pesticides and fertilizers and protection of soil and water resources. Biodiversity was a major focus, and considerable hours in interviews and on questionnaire surveys were spent discussing how farmers protect or fell shade trees, hunt wildlife or destroy habitats. On occasion, I pestered farmers to show me around their estate, draw maps, and explain over and over aspects of their everyday reality that they must have found preposterously trite. So why did they agree to meet me again and again?

The coffee landscape in India is visibly dominated by men. The post-colonial plantation culture involves managing dozens of plantation laborers, procuring and applying barrels of chemical sprays and fertilizers, organizing the lopping of branches of tall trees, strategizing on yearly coffee yields, and negotiating the market for coffee sales. All of these activities are seen as too 'tiring' or 'tricky' for women farmers, especially since all of this involves engaging predominantly with other men. The role of women in this landscape is primarily in domestic support. On occasion, women take up part-time work as school teachers. From my experience, the day-to-day decision-making presence of women in plantation activities is minimal.

The non-visible role of women in plantations meant that my interactions were almost exclusively with men, but this was not so straightforward. The region has an undeniable atmosphere of machismo, perhaps a remnant of the region's strong military history (in the colonial British forces and later in the Indian armed forces) and even earlier references to the region's residents as a 'warrior race' or 'gallant tribes' living in 'remote, highly forested areas amongst wild beasts.' Traditions of game hunting were strong until 1972 when Indian legislation prohibited the hunting of protected wildlife species. Residents enjoy rights to possess firearms without requiring official approval from the state, the only such legal privilege to be allowed to anyone in India. At functions, men wear *kupyas* (black knee-length half-sleeved coats). A maroon and gold sash is tied at the waist with an ornate silver dagger, known as the *peechekath*, tucked in. The *odikathi* is another knife tucked in at the back and a small gun and dagger is hung on to the sash to complete the martial look. My interviewees frequently picked me up in a noisy jeep wearing muddy gumboots and khaki hats with a rifle resting on the backseat – nothing short of a tryst with an Indian Indiana Jones!

The context of masculinity is important because it framed the context of my positionality as a female researcher. Partially smitten myself, partially pragmatic, within the first week I knew that being a woman was going to matter. Research has been a game of strategic diplomacy and having precariously play-acted this during my years of fieldwork, I am convinced that diplomatic success is linked to far more than just the content. I argue that performing 'woman' matters, much more than accounted for in academic discussions on research ethics and methodology.

What happened in the field

Eszter, Hungary

Deciding to study land use politics meant that I had to engage with the Hungarian state's Agriculture and Rural Development Agency, which at the time had taken a decidedly secretive turn. Participation rates in agri-environmental schemes (which were funded publicly through European Union and national monies) were not easily or transparently available, and I encountered a work climate where most people feared for their jobs and were not comfortable sharing information or going on the record. Due to this I was reliant on personal contacts and help for a literal foot-in-the-door. The gatekeepers at both a policy-making level and 'in the field' (in local-level farming contexts) were male; they included the heads of farming families, leaders of hunting associations, farm administrators, and National Park Rangers.

I do not wish to paint a picture that the conservation sector in particular, nor any other sector, is entirely sexist, but I experienced that being a girl in the initial stages of fieldwork was important. I certainly became, or was made to be, hyper-aware of my gender; I began to see how I could use it as a play-thing to get what I needed. Twice in exchange for interviews I fell into a flirtatious game of being asked, 'If I talk to you, what do I get?' and whereby my future interviewee requested – seemingly kindly and frankly – that I wear a short skirt to our interview. After recognizing the theater that some men expected of me, I frequently played nice. Essentially I grinned and bore it. I played the role that would allow me to progress with my work, but through doing so would do nothing for progressing equal treatment for women. Apart from these rather overt propositions, there were several cases of what I would call 'polite' sexism, where praise for one's work was mixed in with reference to gender and relationship status, and an assumption that by just being there, alive, interested in asking the man questions, and (hopefully) single, I was looking for a husband, and that naturally such attention from my side was not unwelcome.[3]

This had interesting ramifications not only on how I was perceived as a non-conforming researcher, but on the research subject as well. I had many conversations where, in response to the question, 'What do you study?' I replied, 'conservation and agricultural politics,' whereupon the retort was something along the lines of 'that's a complicated subject for a girl.' Some interviewees assumed that this meant I was more interested in conserving cute animals than understanding the rules and effects of applying chemical limitations to farmland, and in such cases it was difficult for me to attain more than superficial answers to these more 'serious' questions. In response to questions on the latter I was sometimes told 'not to worry about it' or 'that's very complicated for you to understand.' Thus, gender stereotyping not only belittled me, but in several instances I encountered difficulties in having my questions taken seriously. I frequently felt ethically compromised and somewhat dirty. When I played along and adopted the expressions and humor that I felt was expected of me, I thereby consented to a form of interaction that I struggle to this day to label as anything other than sexual harassment.

Arshiya, India

Gaining access to people and information was surprisingly easy for me. Each coffee grower willingly, courteously offered to introduce me to another and another. Very quickly, a target of 30 farmers snowballed to over 90 in the course of just a few months.

This story is two-fold. The 'let's take it at face value' explanation is that gaining access was easy because I am a nice person. The 'juicier' explanation is that there was underlying sexuality because I was consciously 'performing' being nice. As a young, female, attractive researcher, I was never perceived as a threat. I performed nice to ease suspicion, offer transparency, and make my position and interest in farmers' lives seem harmless and believable. I worried about how I looked: accentuated my eyes with *kohl* (a traditional cosmetic used to darken eyelids) and discarded my spectacles (remembering Malcolm Gladwell 2005 on how good-looking people have a higher chance of successful outcomes).

Given the high education level, exposure, and outspokenness of farmers, I became keenly aware that my access to information would depend entirely on how much my respondents liked me. After repeated pestering for interviews, coffee growers would agree to meet me only if they enjoyed my company, and so I willingly adorned the avatar of the nice 'Good Indian Girl' and practiced the art of being charismatic rather than adopting more formal interview protocol. With this as the starting point for every interview or engagement with a respondent, the possibility of an objective situation was already compromised.

From a conventional research ethics perspective, my performance might be viewed as controversial or questionable or be criticized for compromising objectivity. However, in order to survive in the real world and get my research done, I felt that I had to establish a personal relationship with people. In my context, it would be rude or unacceptable to extract responses to questions and leave. Like many of the other contributors to this volume, I felt 'obliged' to 'give back' (see especially the chapter by Staddon). As a graduate student, I was in no position to promise solutions to complex environmental and development problems, but I could give back in the form of a two-way relationship (see also Godbole in this volume). I frequently provided information about myself and engaged in casual banter on topics of interest to the interviewee. Developing such relationships with respondents is not uncommon practice for researchers but my experience was that in predominantly male scenarios, the chemistry was established far more quickly. Casual banter, charm, and a mildly flirtatious air drew me closer to my respondents such that they liked me and wanted to meet me again. My mannerism, *my sexuality*, was my entry point.

I adopted gender roles that were expected of me by performing the nice researcher and Good Indian Girl, and I pandered to the male ego by silently engaging in tedious monologues on topics that were not relevant to my research or not of personal interest. I stuck to these roles because I did not have a point to prove otherwise (I am fairly secure in my otherwise independent and capable personality). Performing the 'distressed damsel' (mostly in relation to having no means of transport!) helped to gain people's sympathy and support, and in return I avoided the risk of elephants when walking five kilometers through deserted coffee plantations in the late evenings. Despite my performances of gender, obvious propositions by men were absent. This was very welcome but not one bit surprising, given the extremely strong family ties and traditions, the close-knit clan structure, and the taboo that any sexual relation with me would have accompanied.

At the same time, it would be unfair to myself to write about my experiences without slumping into my chair, sighing, and saying that performing sexuality has been exhausting. It has been physically and emotionally draining to be con-tinuously charismatic. Over time, I began to feel as if I was using myself or com-promising my integrity. Quite frankly, I am a naturally flirtatious woman. I am good at conversational banter, making people feel at ease, and breaking the ice. However, the realization that I was flirting strategically left a rather bitter taste in my mouth. It took me a year to realize that fundamentally I was uncomfortable, and so I stopped.

Discarding this once-performed sexuality had repercussions for how keenly my respondents engaged with me. The moment that sexuality was removed from the relationship (let's say I mentioned I was in a long-term relationship), my male respondents would withdraw. They were less enthusiastic about meetings, left me to travel alone, and overall behaved with more restraint and formality. I had not necessarily seen my relationships with respondents as sexual until I com-pared them with 'lack of sexual.' The fact that performing sexuality in the field has *trade-offs* was very much a hard reality.

Walking ethical tightropes

We have written from two dramatically different contexts. In Hungary perform-ances of sexuality and gender were overt, but in India they were more subtle. However, in both experiences we came to appreciate the importance of blending in, of finding one's 'place' and acceptance among a community as well as within the confines of a one-on-one interview. The role of gender in conforming, finding place, and shaping boundaries is what we have wanted to explore through recounting our experiences (Besio 2003) (see also Le Masson in this volume for a discussion on the role of gender in shaping access to informants and the data collection process).

Gendered relations in our work were pivotal for how we gained access to respondents, but instead of considering this as a passive, unavoidable by-product of being human, researchers need to acknowledge that they can potentially *perform* their gender as leverage or play out charm to gain more access. The ethical question is whether a researcher should actively do so (see also the chapters by Dam and Lunn and by Godbole in this volume on deliberately manipulating other aspects of identity).

There are a million 'shades of gray' in the sexuality of relationships; while we are not in any way advocating actual sex as a way of gaining trust, empathy, or access, we are recognizing and admitting that sexuality did exist in our emerging relations with respondents. Whether we encouraged our respondents to feel protective and sympathetic or performed gender to heighten our rapport with our interviewees, these dynamics between researcher and subject do need to be acknowledged and debated. As researchers, we represent educational institutions, other researchers, and ourselves. That is, we are not talking about being dishonest vis-à-vis our research respondents. Rather, how dishonest or muted are our research findings if we do not reveal how we came to gain them? Forget subtleties: if research participants are willing to exchange sex for access, would it be ethically 'wrong'? Should it be admitted to? At the very least, shouldn't we be discussing it?

The issue of positionality in social science research has always been uncomfortably associated and tied to the need to remain 'objective.' It might be useful to conceptualize different 'scales' of objectivity, where social boundaries are mutually negotiated and allow for information mining. This would only be compromised if the quality of the information is in some way affected by the standing and relationship that exists between the researched and researcher. Naturally, our ability to assess the quality of information may be more compromised when we are closer to our research informants, and this boundary deserves closer scrutiny and reflection.

Regardless of the country, culture, or social network, relationships can become sexualized. We have shared our stories from two very disconnected cultures but we have both felt that our positionality as women morphed into more than what we were prepared for as doctoral researchers. Our research was situated in largely patriarchal contexts where women were not major gatekeepers of information. In other cultures, the interaction between women researchers and women gatekeepers would have entirely different sexual dynamics.

The truth is that we work in difficult situations wherein we are continuously dependent on people for their time and help. Our fieldwork is dominated by pleas for interviews and the struggle to get access to information. From both a pragmatic and equitable perspective, we feel that it has been important to engage respondents in our research and make them feel a part of the process. In this respect, our 'fields' have been an extension of our human nature, our lives, our relationships (Kindon and Cupples 2003). We have worked in isolation and lived in unfamiliar places without friends or family, and we have sought connections with our research subjects. The people we interviewed were not the 'other.' They were as we were and we sought to enter their worlds to gain understandings. Our relations made us human and the need for these connections persists even in the field and on the job. We take our own biases and behavioral expectations where we go, and these must interact and find their place among the 'researched.'

Learning to cope, knowing exactly what to do and how to handle a given situation, cannot be taught or learnt from a textbook. We are too founded as *individuals* to be able to judge each other on what is ethically or morally appropriate. Our subjectivities and positionalities as researchers make us vulnerable, whether with our respondents or ourselves. This tightrope must be personally negotiated. However, what we can do is be open to recognize our frailties and moments of self-doubt and share our stories with fellow researchers.

Acknowledgments

Eszter would like to thank Àgnes Kaloczkai, Bàlint Balàzs, and Jànos Gyüre for constructive and helpful comments on this paper.

Arshiya would like to thank those, in the field and at home, who ensured that despite many awkward moments and close calls, she was always safe.

Recommended reading

Butler, J. (1990) *Gender trouble: Feminism and the subversion of identity*, New York: Routledge.

This work examines and highlights how gender is, involuntarily, socially and culturally constructed and performed through repetition and expectation.

Cupples, J. (2002) 'The field as a landscape of desire: Sex and sexuality in geographical fieldwork,' *Area*, 34: 382–390.

This interesting paper explores the impact of sex and sexuality on cross-cultural fieldwork and calls for greater self-reflexivity.

Kulick, D. and Willson, M. (eds) (1996) *Taboo: Sex, identity and erotic subjectivity in anthropological fieldwork*, New York: Routledge.

This collection presents the insights and experiences of a range of ethnographers, addressing how the sexuality of researchers and their research may influence the production of knowledge.

Moser, S. (2008) 'Personality: A new positionality?,' *Area*, 40: 383–392.

Moser describes the profound ability of personality to influence the research process.

Pini, B. (2004) 'On being a nice country girl and an academic feminist: Using reflexivity in rural social research,' *Journal of Rural Studies*, 20 (2): 169–179.

This paper explores notions of feminine identity and different roles that women adopt.

Notes

1 These comments were in response to a motion to pass a bill for the protection of women suffering from domestic violence. The speech went on to say,

> [women] should be focusing on not giving birth to one or two children for our society, but three, four or five. And then mutual respect would make sense within the home, and domestic violence, rape within the family, would never occur.

A significant outcome of these comments was actually such widespread criticism and public pressure that the bill passed.

2 I would like to again stress here that I have in mind three experiences and incidents that I came across in three different National Parks. I am not alleging in any way that all Hungarian National Parks suffer from an over-abundance of self-ruling alpha-males, or even that overt sexism is the norm. On the contrary, I believe that quite specific conditions in these cases created the social dynamics and hierarchies that I observed, which include isolation, absence of or a lack of partners and long-term relationships, a shortage of women in rural areas that led to tight male bonds, and the like. I am describing *my* personal experience in the field, and do not wish this to be confused with an overall 'snapshot summary' of the state of affairs across Hungarian National Parks.

3 A colleague and good friend, upon reviewing a draft of this text for me, commented on this section, 'This is true of all men! Not just in your work! All men are flattered and hope their attentions are welcome!' For me, this comment belies the difficulty of writing about sex and flirtation 'in the field,' as it presumes a separation from our everyday selves and from the 'typical' interactions that may occur between men and women, not just interviewers and interviewees. But this is the key aim of this chapter: to emphasise how relations in our everyday lives do not cease to affect our fieldwork techniques, even though we inscribe an effort to be emotionally un-engaged, to remain 'objective.'

References

Besio, K. (2003) 'Steppin' in it: Postcoloniality in northern Pakistan,' *Area*, 35 (1): 24–33.

Butler, J. (1990) *Gender trouble: Feminism and the subversion of identity*, New York: Routledge.

Gladwell, M. (2005) *Blink: The power of thinking without thinking*, New York: Little, Brown.

Kindon, S. and Cupples, J. (2003) 'Anything to declare? The politics and practicalities of leaving the field,' in R. Scheyvens and D. Storey (eds), *Development fieldwork: A practical guide*, London and Thousand Oaks, CA: Sage.

Kulick, D. and Willson, M. (eds) (1996) *Taboo: Sex, identity and erotic subjectivity in anthropological fieldwork*, New York: Routledge.

Metcalfe, B. D. and Afanassieva, M. (2005) 'Gender, work and equal opportunities in central and eastern Europe,' *Women in Management Review*, 20 (6): 397–411.

Pollert, A. (2003) 'Women, work and equal opportunities in post-communist transition,' *Work, Employment and Society*, 17 (2): 331–357.

Weiner, E. (2009) 'Eastern houses, western bricks? (Re)constructing gender sensibilities in the European Union's eastward enlargement,' *Social Politics*, 16 (3): 303–326.

11 Family connections: ethical implications of involving relatives in field research

Experiences from fieldwork on squatters and land rights in Jamaica

Luke Taylor

Yes sir. You know the Taylors from Mitchells Hill?... That's right! This is Mr Taylor's nephew. He's visiting to talk with you about some research he's doing. You have time fe talk?

This was typically how my Jamaican uncle introduced me to residents of Mineral Heights, an illegal squatter community in Clarendon Parish. My relatives played a significant role in my field research in Jamaica in a variety of ways, both before and during my time in the country. Their advice, local knowledge, contacts, and physical presence enabled me to carry out my research with attention to certain ethical concerns raised by the context in which I was working. However, their very involvement in my fieldwork raised another set of ethical questions about involving relatives in research. These are the issues discussed in this chapter.

An increasing number of researchers in development studies choose to conduct fieldwork at 'home' (i.e., their country of origin) or in a location where they have a close personal connection (see, for example, Chopra, Dam and Lunn, Godbole, Kovàcs and Bose, Leck, Perera-Mubarak, and Wang in this volume). I am of dual heritage – English and Jamaican. Although I grew up in the UK, I have visited Jamaica at regular intervals throughout my life and have a particular interest in issues relating to the country's socio-economic development. During my postgraduate studies in international development and planning, Jamaica was my natural choice of field location: I could combine my personal interest in the country with my favored research topics of the built environment, social rights, and community empowerment.

Despite the personal connection with Jamaica, I was not an 'insider.' I knew that my field research would depend on key gatekeepers who would enable me to gain access to the communities that I wished to study. This access was a particular challenge as I wanted to study illegal squatter settlements, by their very nature comprised of people living at the margins of society, partly hidden and also vulnerable. In addition to this, my fieldwork was conducted at a time when there had been political tensions and violence in some squatter settlements. Thus finding the most appropriate gatekeepers into these communities was critical to my research.

My family members who lived in the area were the natural choice to be those gatekeepers. My father grew up in relatively poor circumstances in rural Jamaica. While he migrated to the UK to find new opportunities, most of his family remained behind. Had I arrived without any prior contacts, my research would have depended on collaborating with a local organization such as a university department or NGO who could broker relationships with the communities I wished to study. Thus I considered it an advantage to have strong links with my fieldwork location. Furthermore, my family had a genuine interest in my work and wanted me to succeed. Put simply, I do not think that I would have had the same safe and successful fieldwork without the assistance of my family members.

In this chapter I examine the ethical issues relating to three aspects of my fieldwork: my choice of fieldwork location in a dangerous setting and the need to ensure my safety; my access to a partly-hidden community and the challenge of gaining access; and collecting data on a sensitive topic from vulnerable people while protecting them from potential harm. In each case I describe how the information, advice, and interventions of my relatives assisted in addressing these ethical issues. While involving relatives in a research project can have many advantages there are broader ethical implications, and I reflect on some of these at the end.

Fieldwork in a dangerous setting

Some fieldwork in the Global South is done in settings which are politically unstable, among people who are socially marginalized, or on topics which are controversial (see, for example, chapters by Brooks, Skinner, and Tomei in this volume on fieldwork in particularly risky settings). My research investigated how squatters felt about living on land illegally. The period when I was planning my field research coincided with a time of heightened sensitivity to security issues in Jamaica's squatter settlements. There had been a request from the USA to extradite a drug lord, Christopher 'Dudus' Coke, nicknamed the 'President.' He was well-known throughout many squatter communities and was something of a role model and father figure who provided help and support to people in return for their respect and allegiance ('Kingston declares state of emergency' 2010). When the police came to arrest 'Dudus' they faced a mighty battle against the squatter communities loyal to him. The violence, bombings and shootings lasted several weeks and resulted in 73 deaths ('As Jamaican drug lord is sentenced' 2012) before he was finally arrested and taken to the USA to face trial.

Having watched these events unfold and seen horrific scenes of violence on international television, the first ethical dilemma that I faced was whether to go ahead with my plans for field research, not only in a country which had just experienced social unrest but also right among the very communities at the heart of the violence (see also Tomei in this volume, whose choice of fieldwork location was also questioned in the planning stages). In this respect, the presence of my family in Jamaica was very useful in helping me to make the most appropriate decision.

Part of my preparation for my fieldwork was to complete the university's risk assessment procedure. My work was considered as being high-risk, partly because of the type of data that I wanted to collect from vulnerable people and partly due to the turbulent circumstances in the country. The university required me to monitor the situation regularly in the time leading up to my fieldwork. One month before my planned departure date, the British Foreign and Common-wealth Office (FCO) issued advice 'against all but essential travel to Kingston.' I also followed the media's coverage of events. At times, the violence was the lead story on various news channels, with live footage of clashes between the police and civilians in Kingston. The images that were shown on television were very graphic and showed the worst of the civil unrest. There was also up-to-date information available on the internet and I found the website of the island's national newspaper, *The Gleaner*, very useful.

Although the FCO advice and the media coverage made me somewhat con-cerned, I was in regular contact with my family members in Jamaica by tele-phone and email. They reassured me that while the situation in the capital city was still unstable, it was much better in Clarendon, where my fieldwork was to be carried out, and no fighting had occurred there. Had there been unrest in the area my family would definitely have objected to my trip. While they wanted me to visit and conduct my research, my personal safety came above anything else. I trusted their judgment and would have followed their recommendations even if that meant changing my plans.

My family continued to keep me informed of the situation throughout the final few weeks of preparation for my fieldwork. I actually placed more value on the information I received from them than on the FCO advice and media reports, which were less relevant to the locality that I was going to. I stuck with my travel plans and research aims; in fact, by the time I departed for Jamaica the warning issued by the FCO had been lowered.

My recommendation to other researchers planning their fieldwork in poten-tially dangerous or volatile locations would be to seek the advice of locals. I cannot overstate the significance of local knowledge and realistic advice from people on the ground whom you can trust. Those doing research at 'home' are at an advantage in this respect because of their ability to capitalize upon contacts and networks to gather detailed information in advance about potential risks. However, you should triangulate this with other sources of information includ-ing FCO travel advice (or equivalent national government advice for those outside the UK). In fact, many travel insurance policies will not cover you if you go against FCO advice; similarly, funding organizations and university risk assessment panels may use the FCO guidance as their most authoritative source of information on the country and situation. Whatever guidance you seek, though, the responsibility for your choice is ultimately yours and you should have a clear conscience that you have weighed up the risks and made the most sensible decision.

Gaining access to marginal communities

I had chosen to do my fieldwork in my family's local area, the parish of Claren-don. After arriving I had an initial meeting with representatives of Clarendon Parish Council, who gave me a list of different squatter communities within the area. I took this list home to my family to seek their opinion as to which were the most suitable for me to choose as my case studies. I valued their opinion since they had a lifetime's knowledge of the local area. We sat around the dinner table and had an informal discussion, which was a great forum for their know-ledge, experiences, and conflicting opinions to be debated. Eventually we selected three communities which combined geographical accessibility, existing contacts, and only moderate crime risk. This contribution of my family to select-ing my research sample was absolutely invaluable and such detailed insight and choices might not have been possible with other types of gatekeepers.

Having selected the communities that I wished to study, my next challenge was to approach them in order to explain my research and persuade people to take part in interviews and focus groups. However, as an 'outsider' and a 'whitey' it would have been very difficult for me to arrive unannounced and unaccompanied in these settlements suddenly wanting to be socially accepted and to ask questions about sensitive topics.

Any outsider entering a community, particularly a hidden, transitory, or illegal community, is likely to be viewed with suspicion. In fact, numerous social researchers have been accused of being government spies (Lee 1993). Since squatting is illegal, squatters live at the margins of society, often keeping a low profile with regard to the authorities, reluctant to talk to outsiders and mistrustful of any outsider. I was about to enter these communities and ask questions about squatting, informality, and crime at a time when these issues had been pushed into the national and international spotlight with the Dudus case. Regardless of my academic affiliation and who I claimed to be, people could have still assumed that I was there for other reasons.

Even the physical act of entering these communities was a personal risk. As Liamputtong (2010: 226) says: 'one who intrudes into private space may pose a threat of risk to actors which fear exposure and sanctions.' Hobbs and Wright (2006) refer to the case of Ken Pryce, a social researcher who was conducting an ethnographic study of street hustling and organized crime in the Caribbean, and was murdered. This is a stark reminder that the researcher's safety should be paramount in dangerous and volatile settings.

My family members were absolutely crucial in enabling me to gain access to my case study communities and ensuring my safety because of their local know-ledge of the history of the squatter settlements and their existing contacts in these communities. The best term to describe my family's role was 'culture broker' (Crist and Escandón-Dominguez 2003; Eide and Allen 2005). They physically accompanied me, introduced me to their contacts, verified my identity, and built the rapport that I needed to explain my research and gain their approval to carry it out.

For example, one of the squatter settlements on the outskirts of May Pen was well known for its high crime and murder rates. Without a local contact it would have been unwise for me to go there alone and unannounced. However, my aunt visited this settlement every week to sell chicken eggs from her farm. My aunt's contact was a local shopkeeper and their long acquaintance was my gateway into the community. I told the shopkeeper about my experience of being in Jamaica, my family connections, and my research topic. She showed a genuine interest in wanting to help me because of the relationship that she already had with my aunt. She thought about which community members would be most useful for me to speak to and then personally introduced me to them. Within half an hour of my arrival, an informal, unplanned focus group had formed and residents shared their experiences and thoughts with me. A potentially risky environment had been turned into a safe environment by the mediating presence of my aunt.

It is important to reflect on what particular qualities made my family members accepted by the squatter communities. My uncle had lived in England and been successful in his career, but he had always planned to retire back to Jamaica because of his love for his homeland. When he returned, he built a house on his family's land where he had grown up, rather than buying a house in a gated community in the up-market parish of Manchester as is common with returning expats. He also remarried, and his new wife was a primary school teacher. They attended the local church and sold produce from their farm. In these and other ways they were a well-known family in the area and had a reputation for working hard and helping others in times of need. It could have been quite different had my uncle and his family been part of the social elite: for example if he were a prominent local politician or a landowner. This different positionality would have affected his relationship with the squatter communities and therefore my prospects for data collection.

I also think it would have been very different had I been relying on non-family members as gatekeepers. The bond of trust, love, and support with my family members was much stronger than any bond that could be built up in a professional context with other people such as a Jamaican researcher from the national university or a fieldworker from an NGO. The importance of trust was magnified in the context of marginal communities and a dangerous situation.

I was very fortunate in the good reputation that my family had and the respectful relationships they enjoyed with the squatter communities. However, to other researchers considering using their family members as gatekeepers I would recommend finding out as much as possible in advance, not only about which people, communities, or organizations that your family have contacts with but also their positionality. If you assume that they have a certain type of relationship with an individual or group and this proves to be different then this could affect your research as well as having implications for your safety.

Researching a sensitive topic among vulnerable people

Sieber and Stanley (1988: 49) define sensitive research as 'studies in which there are potential consequences or implications, either directly for the participants in the research or for the class of individuals represented by the research.' Lee (1993: 4) considers this to be too broad a statement, as almost any science research has consequences; instead he suggests that sensitive research can be defined as 'research which potentially poses a substantial threat to those who are or have been involved in it.'

Based on this definition, my research was definitely on a sensitive topic. Tenure security can involve the illegal occupation of land, disputed claims over rights to land, and violent enforcement of land rights. It spans the legal, political, and social spheres. Squatters are vulnerable, both as individuals and as communities, because they are living a precarious existence. At any time they could be forcibly removed and their homes destroyed by the authorities or land owners. This puts them in a constant state of watchfulness and fear. There were also potential risks to the participants of being involved in my research should my data fall into the wrong hands and threaten their security to stay on the land (see also the chapters by Brooks, Fagerholm, and Skinner in this volume for examples of similar risks).

A researcher's very act of asking questions draws thoughts and attention to the topic; even just the researcher's presence can offend or worry people (Lee 1993). Zetter and de Souza's (2000) reflexive account of researching tenure security in Recife, Brazil includes incidents of participants losing sleep through worry because of their involvement in the research. Furthermore, vulnerable people who have a fear of anyone in a position of authority and power can be reluctant to divulge information, particularly to someone who writes all their responses in a notebook or records their answers on a machine.

I think that the presence of my family members helped toward mitigating any unease about my presence or any possible concerns about taking part in the research. As discussed in the previous section, using my relatives as gatekeepers to access settlements depended on them having pre-existing contacts within the settlements and being known as people who could be trusted. My family members had verified my identity and reassured community members of the genuineness of my being there. This personal introduction by my family members who were known there changed my positionality so that I become a part-insider. In the case of the local shopkeeper mentioned earlier, the trust that she had developed with my aunt over time was effectively transferred to me. Had I been using other types of gatekeepers it might have taken much more time for them to build a relationship of trust with the communities in order for them to agree to participate in my research.

As recommended by ethical guidelines, I always carried a formal letter from my university that confirmed who I was and the purposes of my research. However, I avoided using it in the squatter settlements because I feared that these people might be mistrustful of an official-looking document (see also Skinner in this volume on varying reactions to informed consent forms). My aunt and uncle had introduced me in an informal way, and my access to the communities was built on this personal connection rather than on a formal procedure. I feel that the genuine interest I showed when listening to people's stories and experiences, together with my body language, was enough of a reassurance that they could trust me with sensitive information. In fact, because of my part-insider status, interviews and focus groups sometimes felt more like an informal chat between community members than a foreign researcher asking questions.

An essential element of working with vulnerable people is to provide assurances that personal and sensitive information will be handled confidentially and responses anonymized in order to protect respondents from harm. It was a key concern for me both when conducting my field research and when writing about it that my participants, their homes, and their livelihoods would in no way or at any time be at risk or threatened as a result of their participation in my research project. Although I made assurances that participants' names would be anonymized in my thesis and not subsequently published, and have made every other attempt to minimize risk, one can never truly know whether one's research has caused any harm in the short, medium, or long term.

Although research on sensitive topics and among vulnerable people involves some extra precautions to ensure an ethical approach (Subedi 2006 and Day in this volume), I would not want to discourage other researchers from investigating such issues. Their study should not be neglected just because of practical or ethical challenges. In fact, Sieber and Stanley (1988) suggest that research on sensitive topics 'addresses some of society's most pressing social issues and policy questions' and 'illuminates the darker corners of society,' so it is all the more important (see also Brooks in this volume, who makes the case for intensive qualitative research on difficult issues). Academic research and policy-driven studies are vitally important for addressing issues such as tenure security, which affects millions of people across the Global South.

Wider ethical dilemmas relating to using family members in field research

Many people make use of family members, friends, acquaintances, and social networks for their field research, but there is a range of ethical dilemmas surrounding this.

First, there is the question of whether it is right to rely heavily on others when conducting a supposedly independent piece of research (see also Le Masson in this volume). In my case, there was an added dimension of whether it was right and responsible to involve them in fieldwork which took them into risky situations where their lives and livelihoods could have been put at risk in the present or the future. When we ask family and friends to help us they are very unlikely to say no, as it is natural human behavior to assist relatives and those close to us. While they may have misgivings about getting involved, they are likely to conceal this for the sake of duty or 'keeping face' (see also Wang in this volume on the significance of reciprocal social relationships). So the ethical question is whether it is right to ask someone to help you when really they don't have much of a free choice.

Another major issue to consider is what role your family members should play in your research. Offering practical advice, local knowledge, or emotional support are background roles, but acting as a gatekeeper or assisting in the data collection are more direct roles (see also the chapters by Le Masson and by Lunn and Moscuzza in this volume on the use of partners in the field). You need to question whether their direct involvement with the research participants might influence the data collection. Part of this is considering their positionality in the research community and how it changes your positionality (again see Le Masson on the advantages of using her partner in a gendered approach to fieldwork). My experiences were that my relatives were an advantage to my research; other researchers may find that family members are actually a barrier to accessing the research participants or threaten the objective collection of data, in which case it may be better not to use them and to seek alternative gatekeepers.

Related to the question of what role relatives play in your research is the issue of how to thank them for their involvement. As noted above, most relatives will get involved out of a sense of familial duty and thus may not expect any financial payment, but you need to recognize that they may have dedicated time and personal resources to assisting and enabling your fieldwork. On the other hand, your relatives may have certain expectations from you following their assistance with your research. It is a challenge to sensitively ascertain what these are and establish expectations that are realistic.

A further consideration is the impact on your relatives once you have left. Will their involvement in the project continue: for example if you need them to ask some follow-up questions to participants, are your relatives willing or able to do this on your behalf? Does their association with your research project affect their status locally, either in a positive or a negative way? In your absence, is it possible that the research participants go to your relatives with expectations of what they might receive in compensation for their participation? Does it put your relatives at risk in any way in the longer term? For example, if the squatter settlements that I studied were subsequently targeted by the authorities, might my relatives be at risk if the communities thought that they were exposed as a result of my research? Long after you have left the scene, your relatives who are in situ remain the visible association with your research project; you need to consider the implications for them in the short, medium, and long term.

I conducted my field research in a dangerous setting among a partially-hidden community, and I collected data from vulnerable people on a sensitive topic. Each of these elements added a layer of complexity to my work and presented a range of practical and ethical challenges, and the assistance of my family was crucial to overcoming these challenges which included ensuring my safety, gaining safe access to research communities, and protecting respondents from harm. Prior to fieldwork I relied on my relatives' assessment of the situation in Jamaica and trusted their advice as to whether or not to travel; their life-long knowledge of the local area was invaluable in helping me to select the most appropriate case study communities; their contacts in the communities were essential in enabling me to safely enter the settlements and meet potential respondents; and their reputation in the local area and the levels of trust they had built up with the communities over the years prepared the way for me to discuss sensitive issues in a context of trust.

Overall, I feel certain that my period of fieldwork and the outcome of my research would have been very different had I used different types of gatekeepers rather than my family members. In addition, my family enjoyed being involved in my research and likewise I was pleased to be able to combine an academic project with my family connections. Although involving my family members was a sensible choice, resulting in a successful piece of research, and was enjoyable, family involvement may not be the most appropriate choice for all researchers conducting research at 'home.' There are many ethical issues to consider before involving them, including being aware of their positionality, clearly defining roles and expectations, assessing potential risks, and considering suitable compensation.

Recommended reading

Boas, M., Jennings, K. M. and Shaw, T. M. (2006) 'Dealing with conflicts and emergency situations,' in V. Desai and R. Potter (eds), *Doing development research*, London: Sage.

This chapter provides first-hand accounts of conducting research within conflict and post-conflict zones and makes suggestions of how research can be conducted in a conflict-sensitive manner.

Lee-Treweek, G. and Linkigle, S. (eds) (2000) *Danger in the field: Risk and ethics in social research*, New York: Routledge.

I particularly recommend Chapter 4 by Janet Jamieson, which focuses on experiences of negotiating danger while conducting fieldwork and includes the importance of planning, team-working, and colleague support in countering physical risk. It allows you to think of possible situations which may arise and how danger can be minimized through adequate preparation.

Liamputtong, P. (2010) *Performing qualitative cross-cultural research*, New York: Cambridge University Press.

This excellent volume covers various issues relating to cross-cultural research including insider/outsider positionality, cultural sensitivity, and reciprocity. Chapter 3 is particularly useful for its clear and concise introduction into thinking about gaining access and its discussion of the term 'culture broker.'

Sriram, C. L., King, J. C., Mertus, J. A., Martin-Ortega, O. and Herman, J. (eds) (2009) *Surviving field research: Working in violent and difficult situations*, Abingdon: Routledge.

A very useful guide for researchers looking at the practical and ethical challenges of conducting qualitative research in difficult circumstances such as in autocratic or uncooperative regimes, with governmental or non-governmental officials, with vulnerable respondents such as victims of war crimes, or indeed with the perpetrators of such atrocities.

References

'As Jamaican drug lord is sentenced, U.S. still silent on massacre' (2012) *The New Yorker*, 8 June, available at: www.newyorker.cm/online/blogs/newsdesk/2012/06/christopher-coke-tivoli-massacre.html (accessed 9 June 2012)

Crist, J. D. and Escandón-Dominguez, S. (2003) 'Identifying, recruiting and sustaining Mexican American community partnerships,' *Journal of Transcultural Nursing*, 14: 276–271.

Eide, P. and Allen, C. B. (2005) 'Recruiting transcultural qualitative research participants: A conceptual model,' *International Journal of Qualitative Methods*, 4 (2): 44–56.

Hobbs, D. and Wright, R. (2006) *The Sage handbook of fieldwork*, London: Sage.

'Kingston declares state of emergency after gangs attack police' (2010) *Guardian*, 25 May, available at: www.guardian.co.uk/world/2010/may/24/gangs-barricade-kingston-jamaica?INTCMP=SRCH (accessed 26 May 2010)

Lee, R. (1993) *Doing research on sensitive topics*, London: Sage.

Liamputtong, P. (2010) *Performing qualitative cross-cultural research*, New York: Cambridge University Press.

Sieber, J. and Stanley, B. (1988) 'Ethical and professional dimensions of socially sensitive research,' *American Psychologist*, 43 (1): 49–55.

Subedi, B. (2006) 'Theorizing a "halfie" researcher's identity in transnational fieldwork,' *International Journal of Qualitative Studies in Education*, 19 (5): 573–593.

Zetter, R. and De Souza, F. (2000) 'Understanding processes of informal housing: appropriate methodological tools for a sensitive research area,' *International Planning Studies* 5 (2): 149–164.

Part III

Ethical issues relating to research methods

12 Fellow traveller or viper in the nest? Negotiating ethics in ethnographic research

Experiences from fieldwork on international volunteering and environmental conservation in Kenya

Margi Bryant

> Volunteers visited local school today, including me though I was just observing. The idea was to discuss environment and conservation with the kids. Don meant to bring some sketches of forest animals, but didn't have time to do them, so brought his mini-laptop with photos on it, and walked round classroom showing pix on tiny screen. Not a good idea – pix were very hard to see, it took ages to get round class, and it was flaunting technology which these kids haven't got. Then we talked about marine conservation. Lucy asked class: 'If you see shells on the beach, is it good to take them?' Several kids replied immediately: 'Yes, it's good, because if we don't have a job we can sell the shells and make money, we can buy things like uniforms for school.' Someone else chipped in: 'And pencils.' But the volunteers weren't having this, they insisted you should leave shells where they are. (I'm not sure how sound this is as a conservation argument – surely empty shells are dead anyway?) Awkward moment at the end when a boy asked if he could say something. We expected some sort of thank-you but he gave a little speech about how poor he was, and asked us for money. Jane [volunteer group leader] told him the best thing was to work hard at school, this would help him get a job and earn money. Very patronizing of course, but I think we all felt embarrassed, and none of us came up with a better response. We left rather precipitately, then off to talk to another class.

This is an extract from the field diary I kept while doing research in Kenya for my master's degree dissertation. (The names are pseudonyms in this and other diary extracts in this chapter.) Like all field diaries, it forms part of the raw material of fieldwork, to be subsequently processed into emerging themes for academic analysis, so it would never normally see the light of day in this form. Looking at it now, it seems full of subjective impressions, value judgments, and personal criticisms, and I'm certainly not proud of it. But the extract, and even more so the challenge of turning it into 'data,' does illustrate very well some of the ethical dilemmas of ethnographic research. Was I complicit in the volunteers' approach and actions? Should I have tried to persuade them not to take the laptop into the classroom, or suggested a way of helping the boy who asked for money? Would my presence as part of their group influence any future interaction I might have with the schoolchildren or their teachers? Is my critical take on the volunteers' approach valid, and would it be amenable to feedback, discussion, and acceptance?

Ethnography is probably best described as not so much a research method, but rather a style of research (Brewer 2000: 11). Its core approach is 'participant observation,' whereby the researcher joins in the everyday lived experience of the people being studied (see also Brooks in this volume). 'Participant observation' is, of course, something of an oxymoron (how can you truly participate if you're observing, or truly observe if you're participating?), but in practice it can mean anything from being a fully involved member of the group to being a somewhat detached hanger-on. But wherever the researcher is positioned along the participation–observation scale, this approach generally involves a range of social interactions amounting to far more than mere observation. Talking and listening loom particularly large (Forsey 2010), but experiences with all the other senses are also important. Ethnography is often nowadays expanded to include semi-structured interviews, focus groups, photography, solicited diaries, and so on (Crang and Cook 2007), but in an ethnographic study these other methods are built around – and influenced by – the engaged presence of the researcher, rather than as self-contained undertakings.

Originating in social anthropology in the early twentieth century, especially the work of Bronislaw Malinowksi, ethnography has fallen in and out of favor over the past 100 years. In a Global South context, it has been widely seen as inseparable from the colonial encounter and the objectifying colonial gaze. One response to this, in European and American anthropology, has been to switch the focus of study to societies and institutions closer to home. But it's important to remember that the early advocates of ethnography were doing something radical, even perhaps deeply ethical, at the time. In his fieldwork in the western Pacific, Malinowksi shared people's daily routines as a way of understanding their worldview, which he presented as valid and rational, thus challenging some of the prevailing theories and assumptions about 'primitive' peoples (Malinowski 1922). The purpose of ethnography is still 'to understand parts of the world more or less as they are experienced and understood in the everyday lives of people who live them out' (Crang and Cook 2007: 1). Nonetheless, ethnography has had a checkered history in international development research. Robert Chambers was particularly skeptical, seeing it as one-way rather than participatory, and too time-intensive compared to the rapid appraisal methods needed for development interventions (Chambers 1983). However, ethnography has recently enjoyed renewed attention in development studies, particularly through the work of such researchers as Norman Long and David Mosse. They argue that ethnography is uniquely valuable for revealing actor perspectives, picking up contradictions and inconsistencies, rethinking first impressions, and recognizing unexpected outcomes, all of which make it highly appropriate for studying the encounters and interactions peculiar to development practice (Long 2001; Mosse 2005).

My research, for both my MA and my PhD, focused on international volunteering and environmental conservation in Kenya. I aimed to explore how the perceptions and practices of international agencies, Kenyan organizations, foreign volunteers, and local communities intersect, relate, or collide when locally-based conservation projects are assisted by groups of foreign volunteers. Ethnography was my method of choice for several reasons. For one thing, it had novelty value. I had come across a number of published studies of international volunteering based on interviews, but relatively few involving an extended period of participant observation. Secondly, I thought that joining a project as a volunteer would give me an obvious mechanism for being there, and make the whole process of access easier than if I were just turning up to do interviews. I wouldn't need any special skills, just willingness to do whatever work volunteers were required to do. But the most important factor in choosing ethnography as my core method was that it would involve me in the everyday routines, processes, roles, relationships, conversations, and conundrums of a volunteer-assisted environmental project. This would have enormous potential for exploring the dynamics of power and knowledge played out through personal interactions. I felt that questionnaires or semi-structured interviews, even if carried out 'in the field' while the project was underway, would show me only a very small part of the picture. In retrospect, I would still hold to that view, but I hugely underestimated the complex issues of transparency, identity, representation, and accountability which came as part of the package. While none of these are unique to ethnography, I would suggest that ethnography involves particular, or particularly acute, ethical dilemmas. This chapter explores how these arose in my fieldwork in Kenya and how I attempted, or failed, to deal with them.

Do they really know what you're up to? Issues of transparency, disclosure and consent

The ethics guidance we receive as postgraduate researchers emphasizes the importance of transparency. This means giving a full and honest account of your intended research to everyone involved, and making sure that people participate on a basis of informed consent, including the right to withdraw from the research at any time. Our research proposals must get ethical approval from our universities before we begin fieldwork, so we follow various expected procedures, often including a prepared statement about our research to be shown to potential participants, and a printed consent form for participants to sign. My statement summarized the purpose of my research, the methods I would use, and what I would do with my findings, along with assurances about anonymity, confidentiality, and the right to withdraw. I managed to avoid using a printed form by pointing out that some of the community participants might be illiterate, so I was allowed to seek consent orally instead. In fact, my major concern was not so much literacy levels as people's seemingly universal wariness about signing forms, and the power differentials embedded in the act of eliciting a signature (see also the chapters by Skinner and Taylor in this volume).

However, I was aware of difficulties in putting these ethical standards into practice right at the outset when negotiating access to the volunteer-assisted projects I hoped to study. For my master's degree research I approached a Kenyan

NGO that recruits and hosts fee-paying foreign volunteers and asked if I could spend some time with them as a volunteer and carry out my research at the same time. The person I dealt with was a British long-term volunteer with a managerial role at the NGO who was quick to agree to my request. But I had no control over whether, or how, my research aims were explained to other staff members or to the volunteers I'd be working alongside, who were mostly unknown quantities at that point. So the only real consent I'd obtained before commencing my fieldwork was that of a single managerial gatekeeper.

My PhD research involved two well-established UK-based NGOs, their Kenyan partners, local community organizations and stakeholder groups, individual local participants, and international volunteers. Negotiating access was thus a more complex process, with multiple gatekeepers and some degree of conferring (although I was never sure how much) among and within the organizations before they agreed to work with me. Individual participants, however, were a different matter. For one of my case studies, I had the opportunity to contact the international volunteers by email a few weeks before the conservation project began, to explain what I hoped to do and request their cooperation. By that time, however, they'd committed themselves to taking part in the project and would almost certainly have booked their flights, so it would have been difficult for them to change their plans. I had no prior contact with individual participants in Kenya for either case study until I got to the project locations.

I delivered my pre-prepared, ethically approved statement as soon as I arrived, at initial meetings of NGO staff, international volunteers, local stakeholders, and local participants, and thereafter to everyone who appeared on the project's radar, as and when I met them. But I am not at all sure that this was adequate. First, there's the issue of how you frame and express your research aims and methods in the course of explaining them to potential participants. Not only is your summary statement much shorter than an academic research proposal, but its style of wording is usually very different. In my case, concepts and phrases such as 'postcolonial,' 'micropolitics,' and 'power differentials' were replaced by a rather blandly-expressed interest in how the ideas and approaches of international volunteering organizations interacted with those of local partners and stakeholders. This may be simply a matter of using the appropriate register for a given audience (Crang and Cook 2007: 41), but when does a change of wording become a deliberate attempt to sanitize or even mislead?

Second, I found explaining the daily realities of ethnographic research even more ethically problematic. In my prepared statement I said that I'd be carrying out 'participant observation,' spending time with the people involved in the conservation project and joining in their everyday activities. I also made it clear that their collaboration in my research was entirely voluntary and that they could withdraw at any time. But did participants fully realize that they might therefore find themselves subject to 24/7 scrutiny, or that anything they said might be taken down and used in evidence? Did they assume I was 'on duty' as a researcher only when I was visibly taking notes, or only within notional working hours? And did they actually feel able to withdraw from my study once it was under way? When we made the school visit described at the beginning of this chapter, the volunteers knew I was there to observe. I even carried a conspicuous notebook. But if any of them had felt, halfway through the session, that their interaction with the children wasn't going very well, would they have told me they wanted to opt out? Could I have written notes on the day's events, omitting one or two people as if they hadn't been there? I still have unresolved concerns about how participants' informed consent can be meaningful in an ethnographic study, and how feasible it is for consent to be withheld or withdrawn by individuals within the participant group.

Insider and/or outsider? Issues of positionality, identity, and critical distance

Issues of positionality permeate all social research methods, but they loom particularly large in ethnography because the researcher is performing an identity, or perhaps more than one identity, for a sustained period of time. At the outset, your choice of position along the sliding scale from detached observation to immersed participation is likely to be influenced by identity indicators such as gender, age, ethnicity, nationality, economic position, education, and so on. During my master's degree research, I was one of several foreign volunteers working for the same Kenyan NGO at the same time. We all shared certain characteristics: we came from higher-income countries, we were well-educated, we were interested in conservation, and we were well-off enough to get ourselves to Kenya and pay money for this volunteering experience. We also all stayed in the same accommodation, sharing evening meals and relaxation time as well as work activities. This made it easy for me to blend into the volunteer group, but it was much more difficult to have a similar relationship with the NGO's Kenyan staff, who went home at the end of the working day and never socialized with the volunteers. The obvious disadvantage of my position was greater separation from the staff's lived experience and worldview, while the obvious advantage was the insight I gained into the assumptions, expectations, and privileges that accompanied the volunteer role. An extract from my field diary illustrates this:

> In the office this morning with Eva [acting manager]. She asked what I thought of their guided 'eco-tour.' I expressed some doubts about the way it's currently done, and mentioned work I've done in the UK on guided walks in national parks etc. I suppose I said this to show I had a bit of history with this sort of thing and wasn't just firing off unfounded criticisms. Eva immediately suggested I run a workshop for staff next week. I made a point of saying I don't want to impose my ideas, but she got straight on the phone to their regular guides and fixed up a date for a workshop. Later, she said she must remember to tell Hakim [Kenyan staff member]: 'We must make sure we tell him, he's their boss.' I had no idea he was line manager for the tour guides! Surely we should have discussed the whole idea with him first? I felt very uncomfortable about this, it seemed like Hakim was being sidelined.

Incidents like this provided rich material for one of my emerging themes, namely the privileging of volunteers' knowledge and agency over that of local staff. It also gave me an insight into the everyday mechanisms, in this case a chat in the office, by which privilege could be granted and accepted, which would not have been easy to discern from interviews. But this kind of incident also raised ethical issues around my complicity in the very assumptions and practices I was trying to examine. I did give the workshop on guided walk techniques based on similar training I'd previously delivered in the UK, and although the Kenyan guides said they found it useful, it added to the toll of instances of authority being unquestioningly accorded to newly-arrived volunteers. But I felt that refusing to do it would have been incompatible with my role as a fully participating volunteer. In retrospect, I feel I may have become too much of an 'insider' in this research, notwithstanding the insights it afforded.

In my subsequent PhD fieldwork, I tried harder to establish a distinct identity as a researcher, not wholly absorbed into the culture and practices of the volunteer cohort. I joined in the daily work activities, evening discussions, all meals, and most social events, but I stayed in local lodgings rather than with the group. This sounds like a relatively small adjustment, but I think it had some effect on how I was perceived by both the volunteers and the local community. I also felt less obliged to consistently conform to the volunteer role. After the volunteers had left, I remained for some time at both project locations, in order to deepen my understanding of the networks and actors involved and the context in which they operated. This meant spending much of the day in places where project-related activities continued year round, without any input from volunteers. I was no longer part of a group, I had no specific daily tasks to perform, and I was much more clearly an outsider. My research style shifted several points along the participation-observation scale, becoming a more detached kind of ethnography based on conversations, occasional tagging-along on activities and sitting in on meetings, and the popular ethnographic research practice of 'hanging out' (Wogan 2004). Obviously this raised a different set of positionality issues, and my worries about over-immersion and complicity were replaced with concerns about misunderstanding and misrepresentation. I hope, however, that the working and social relationships I'd already established locally helped to mitigate the worst excesses of the outsider's gaze.

Among these shifting positions and identities, it can be hard to keep track of 'the real you.' Personality, alongside positionality, may play an under-recognized role in fieldwork (Moser 2008), but in ethnographic research you may feel as though you've put your personality and personal feelings on hold (see also Smith in this volume on the role of personality and emotions in interviewing). While this may seem primarily a psychological conundrum, it has ethical implications too. Whether researchers should reveal their own perceptions and opinions, and what happens when they do, are contentious issues in all qualitative research, but particularly in ethnography, where interactions can be intensely personal and can involve wide-ranging informal conversations (Keith 1992). In the course of my fieldwork, I sometimes found myself at odds with fellow volunteers' views and/or behavior, but kept quiet in the interests of maintaining cordial relations and not jeopardizing subsequent research. As someone who normally expresses her opinions quite freely, I found this an uncomfortable experience, as though I was maintaining critical distance not only from my research participants but from myself. Had I been an 'ordinary' volunteer with no other agenda, I would have been opinionated and proactive; as a researcher, I was reticent and passive.

While I don't subscribe to the illusion that researchers can ever be a disembodied, quasi-invisible presence (c.f. Haraway 1991: 189), there have certainly been instances in my fieldwork when I consciously chose to minimize my influence on events. For example, in the school visit described earlier, the 'real me' would have tried to be more actively involved in planning the visit beforehand, suggesting that we used printed pictures instead of a laptop and that we didn't impose 'right' and 'wrong' answers. But the 'researcher me' wanted to know what the other volunteers would do. In my later PhD fieldwork, I've been more upfront in expressing my personal opinions in informal conversations, but have still tried to avoid directly influencing what takes place.

In some circumstances, suppressing personal perceptions may be downright unethical. If they emerge later on in your analysis and interpretation, earlier reticence may come to look like duplicity, and your research participants may see you as a 'viper' unobtrusively ensconced in the research setting, awaiting your moment to strike (Crang and Cook 2007: 27). Nancy Scheper-Hughes' account of the hostility she faced when she returned to her research location in Ireland, after the villagers became aware of her published monograph on their community, is a cautionary tale in this respect (Scheper-Hughes 2000). In the villagers' eyes, the amiable researcher with whom they had shared their lives, thoughts, and secrets had turned on them and 'run them down.'

Whose version of reality? Issues of feedback and authentication

In recent years, the process of discussing preliminary findings with your research participants, and taking their responses into account when forming final conclusions, has come to be seen as a key element of ethically sound research (see, for example, Fagerholm in this volume). It can help to adjust the balance of power, traditionally vested in the researcher, and to correct misinterpretations or misunderstandings on the researcher's part. The Royal Geographical Society/IBG's Developing Areas Research Group (DARG), a prominent source of guidance for fieldwork in the Global South, urges researchers to consult principal stakeholders and representative participants and to strive to reach agreement on findings and their interpretation (DARG 2009). But in many research situations, this process is fraught with difficulties. Contradictions, divergences, and inconsistencies, which are important research findings in themselves, may be ironed out in the consultation process; power differentials among participating individuals and groups can make it impossible to secure meaningful endorsement; and powerful participants can in some circumstances effectively censor research findings (Bradshaw 2001). These problems can arise regardless of what research methods are used, but they are particularly acute when the researcher's approach is highly interpretive, as it is in ethnography. One of the key features of ethnography is its ability to examine the gap between what people do and what they say they do (Herbert 2000). If the researcher wants to authenticate preliminary findings through consultation with study participants, and to reach agreement on interpretation, does this mean that the participants have to recognize and admit that gap? Trying to persuade them to do this can be socially transgressive, confrontational, and ultimately counterproductive (c.f. Mosse 2006).

One of my case study projects emphasized, in all its recruitment and briefing material, the close collaboration between international volunteers and the local community in the village where the project was based. On the first day, the newly-arrived volunteer cohort was taken on a conducted tour of the village by a staff member from the Kenyan implementing organization, who was a long-term local resident. My field diary records my experience of that tour:

> Village tour this afternoon. We set off in a group led by Hassan. We so look like tourists, everyone except me has about £1,000 worth of photographic kit slung round their necks. I'm clutching my notebook. Hassan leads us along lanes and round the houses, pointing out interesting things: different building materials reflecting how well off people are, the old (manual) and new (electric) village pumps, the village school etc. The school wall is adorned with names of sponsors, and we're told how a previous volunteer gave a lot of money towards it. I asked if this was her own money or through fund-raising, but he's not sure. I get the feeling this story is 'pour encourager les autres,' i.e., us! We stop frequently at the volunteers' request so they can take photos, mostly of people. We've been told [in project briefing material] always to ask people's permission to take pix of them. At first one or two of the group ask Hassan as an intermediary ('Is it OK if I take her picture?'), but he says it's fine, and after a while no-one bothers to ask. There are frequent stops for a passing child to be grabbed and posed with, especially by a couple of female volunteers who seem to be channelling Angelina Jolie! We have virtually no interaction with adult villagers. Hassan seems to know everyone in the village, he smiles and nods at people as we pass by, but he doesn't introduce us to anyone. I find the whole tour a deeply uncomfortable experience – it feels impersonal, intrusive, very much like 'performing' tourism. But we're not supposed to be tourists, we're supposed to be collaborators in a community-based project!

In my subsequent conversations with the volunteers, none of them shared my perception of the village tour as an uncomfortable experience, but generally saw it as a good introduction to the local setting. In my subsequent conversations with villagers, many of them felt the volunteers were kept too separate from local residents, and said that they would have liked more interaction. This was meant as a general assessment of the volunteers' presence and activities, not just as a verdict on the village tour, but it was reassuring to find that it echoed my subjective experience. This divergence of perception between volunteers and local residents became a key ingredient of one of my main themes: the gap between rhetoric and reality in 'community-based' conservation volunteering.

But where does that leave the process of authentication? As David Mosse points out, ethnographic analysis is a 'positioned interpretation' which does not rule out other versions (Mosse 2006: 941). With no obvious formula for consensus among different interest groups, it seemed to me that I could either ask each group separately to 'authenticate' what they had told me in the first place, which seemed a somewhat pointless exercise, or I could attempt to raise awareness across the board of the various groups' differing perceptions. The latter was the only way that I would in any meaningful sense be sharing my research findings, but of course it was easier said than done. Some of the views expressed to me were highly critical of other organizational and individual actors, and power differentials among various interests in the project made me anxious about stirring up trouble, however carefully I anonymized individuals. On the other hand, I was aware that in some circumstances, people sharing their views with me fully intended these views to be conveyed to other actors, especially to more powerful interests where I could serve as a useful conduit. The process was also complicated by the inevitable time-lags between various stages of fieldwork and data analysis, and the short-term nature of the volunteers' involvement. So the issues of feedback and authentication for this case study needed to be approached both diplomatically and strategically. Whatever the expected ethical practice, there is also a strong ethical argument for considering possible negative consequences, and it seems reasonable that authentication should always be modified by responsibility.

Ethical or moral? Some concluding thoughts

Ethnography 'uniquely explores lived experience in all its richness and complexity' (Herbert 2000: 551), but it also poses particular ethical challenges. The features that make it an especially productive method – enabling researchers to immerse themselves in a group or institution, engage with actor perspectives, uncover discrepancies between words and actions, and gain 'unreplicable insight' (ibid.: 550) – are the very things that can make it an ethical minefield. Established frameworks and guidelines seem both unhelpfully abstract, using terms like 'integrity, quality and transparency' (ESRC 2012: 2), and inappropriately prescriptive, with procedures such as printed consent forms. In my ethnographic fieldwork in Kenya, I found that the ethics training and ethics review process I'd previously undergone were of very little help with day-to-day dilemmas. My experiences raised a number of questions that I found hard to answer. Are the ethical standards we're expected to meet not merely impossible but actually inappropriate? Are these standards based on incommensurate priorities, namely, individual rights and collective benefits? Are we making assumptions about power relations between the researcher and researched which do not reflect the particularities of fieldwork contexts? Are we disempowering research participants by our paternalistic determination to protect them from 'risk'? And are we thus undermining the capacity of our research to have real value in the world beyond the academy?

Similar questions have been raised in recent critical geography literature. One fundamental problem is that frameworks for ethical guidance in the social sciences have been imported from biomedical research without due regard for the major differences, both epistemological and practical, between these two kinds of endeavor. Dyer and Demeritt (2009) point out that the excessive focus on protecting the rights of individual subjects may obscure or subvert wider moral considerations such as freedom of inquiry and the duty to expose injustice (ibid: 60). Ethnographic research is mentioned by these authors as a particular instance where procedures such as signed consent forms make little sense, yet they point out that the Economic and Social Research Council's (ESRC 2012) research ethics framework makes no allowance for ethnography to be an exception. Research ethics governance thus becomes 'a process of ticking boxes ... rather than making substantive ethical judgments' (ibid.: 59). Gill Valentine, in her very useful *Progress in Human Geography* review (2005), contrasts the 'rubber-stamp' approach of ethics governance with the far deeper considerations of human geography's 'moral turn.' New researchers therefore need better preparation to be able to develop 'individual responses to potentially unique circumstances' (Valentine 2005: 485).

This sounds very familiar to anyone who has done ethnographic research, where every circumstance feels unique and, however close you are to your research participants, negotiating ethical conundrums is ultimately your own responsibility. But what should better preparation involve? I personally found discussions with other postgraduates, sometimes spontaneous, sometimes in student-organized workshops, but always with a prevailing ethos of honesty and confidentiality, far more helpful than formal 'ethics training.' I also learned a lot from the fairly rare examples of literature that talk frankly about 'real-life' ethical dilemmas encountered in the field and raise fundamental issues of moral responsibility in social research (e.g., Ferdinand *et al.* 2007). If an ethical code is a set of principles governing conduct, it's worth remembering that moral responsibility underlies and informs, and may often unsettle or challenge, that code.

Recommended reading

Crang, M. and Cook, I. (2007) *Doing ethnographies*, London: Sage.

My favorite textbook on ethnographic research, this is a much revised and expanded version of the authors' 1995 publication. Engagingly written, it covers a range of methods, using an expanded definition of ethnography as 'participant observation plus,' and includes a helpful discussion of ethics.

Ferdinand, J., Pearson, G., Rowe, M. and Worthington, F. (2007) 'A different kind of ethics,' *Ethnography*, 8 (4): 519–543.

Four 'tales from the field,' along with analysis and discussion, exploring real-life dilemmas faced by researchers, and highlighting the difference between research ethics and moral responsibility. None of the examples are in the Global South, but it is nonetheless a very instructive read.

Mosse, D. (2006) 'Anti-social anthropology? Objectivity, objection, and the ethnography of public policy and professional communities,' *Journal of the Royal Anthropological Institute*, 12 (4): 935–956.

David Mosse's searingly honest account of the controversy surrounding his long-term ethnographic study of a development project in India. Closely analyzes the negative reactions of some research participants in order to explore issues of power, subjectivity, interpretation, and authentication in ethnographic research.

Scheper-Hughes, N. (2000) 'Ire in Ireland,' *Ethnography*, 1 (1): 117–140.

Another reflection on a controversial outcome, describing the author's return visit to her research location in Ireland, when she was confronted with the villagers' outrage at her betrayal (as they saw it) of their trust. Uncomfortable reading but perhaps a cautionary tale.

References

Bradshaw, M. (2001) 'Contracts and member checks in qualitative research in human geography: Reason for caution?,' *Area*, 3 (2): 202–211.
Brewer, J. D. (2000) *Ethnography*, Buckingham: Open University Press.
Chambers, R. (1983) *Rural development: Putting the last first*, Harlow: Longman.
Crang, M. and Cook, I. (2007) *Doing ethnographies*, London: Sage.
DARG (Developing Areas Research Group) (2009) *DARG ethical guidelines*, available at: www.devgeorg.org.uk/?page_id=799 (accessed 1 Dec 2012).
Dyer, S. and Demeritt, D. (2009) 'Un-ethical review? Why it is wrong to apply the medical model of research governance to human geography,' *Progress in human geography*, 33 (1): 46–64.
ESRC (Economic and Social Research Council) (2012) *Framework for research ethics 2010 (revised September 2012)*, Swindon: ESRC.
Ferdinand, J., Pearson, G., Rowe, M. and Worthington, F. (2007) 'A different kind of ethics,' *Ethnography*, 8 (4): 519–543.
Forsey, M. G. (2010) 'Ethnography as participant listening,' *Ethnography*, 11 (4): 558–572.
Haraway, D. J. (1991) *Simians, cyborgs and women: The reinvention of nature*, London: Free Association Books.
Herbert, S. (2000) 'For ethnography,' *Progress in Human Geography*, 24 (4): 550–568.
Keith, M. (1992) 'Angry writing: (Re)presenting the unethical world of the ethnographer,' *Environment and Planning D: Society and Space*, 10: 551–568.

Long, N. (2001) *Development sociology: Actor-oriented perspectives*, London: Routledge.

Malinowksi, B. (1922) *Argonauts of the Western Pacific*, London: Routledge.

Moser, S. (2008) 'Personality: A new positionality?,' *Area* 40 (3): 383–392.

Mosse, D. (2005) *Cultivating development: An ethnography of aid policy and practice*, London: Pluto Press.

Mosse, D. (2006) 'Anti-social anthropology? Objectivity, objection, and the ethnography of public policy and professional communities,' *Journal of the Royal Anthropological Institute*, 12 (4): 935–956.

Scheper-Hughes, N. (2000) 'Ire in Ireland,' *Ethnography*, 1 (1): 117–140.

Valentine, G. (2005) 'Geography and ethics: moral geographies? Ethical commitment in research and teaching,' *Progress in Human Geography*, 29 (4): 483–487.

Wogan, P. (2004) 'Deep hanging-out: Reflections on fieldwork Andean ethnography,' *Identities: Global Studies in Culture and Power*, 11 (1): 129–139.

13 Unsettling the ethical interviewer: emotions, personality, and the interview

Experiences from fieldwork on environmental education in Tanzania

Thomas Aneurin Smith

'I'm not happy with the migration situation. It is high time that the government move from socialism to capitalism. I blame socialism for these problems. I had a farm here. The leaders here told me to give the land to others ... But if there was capitalism I could have had a big farm and I could be making money. So social-ism and the government caused these problems ... If people have money they could use solar power but it seems to be very costly. We can use coal but to use it, well it seems very flammable, so we should get experts to show us how. In 1959 it did not rain for two years but one river never dried. So if this river never dries out how is it possible that we do not have electricity? The government should use the waterfalls for hydroelectric power.'

To be honest I am getting a bit fed up with his [the interviewee's] quite arrogant attitude. I ask David (my research assistant and translator): 'Why does he think they don't have hydroelectric power then?' David translates, and I sense he approves of the fact that I have finally interjected the rambling blame directed at the government, socialism, experts, (whoever!) as well as the suggestion of hydro-electricity which seems, to me, to be out of touch with local needs.

The man replies: 'The main reason is because of bureaucracy, and possibly because we don't have any MPs who are influential in the government from this area....' And so it continues.

The extract from my interview notes above illustrates how, during a 45-minute interview with a male farmer in rural Tanzania, I developed a distinct feeling of dislike for the participant. I felt as if the interviewee had been patronizing and talking down to me, and that he was utilizing his position as an authority figure in the community in order to do so. In return, this made me feel frustrated, angry, and in many ways 'negative' toward him. This attitude may not seem completely apparent in the words he spoke, yet I still had this distinct feeling based on what he said, how he said it, and how he related to David and me. Such an emotional reaction from the interviewer may seem inappropriate, unscholarly, and unethi-cal. In this chapter I will examine this confluence of emotion, personality, ethics, and the interview.

The interview took place as part of a doctoral research project on local environmental education in Tanzania. A significant part of my fieldwork involved interviewing local residents across three field sites about their perceptions of local environmental problems, and the incident above came from an interview in one of these villages. As a rule, the residents of the village held what might be considered 'traditional' values, particularly in terms of gender roles, and men dominated decision-making in families. Older men in particular were highly respected, and tended to dictate the public affairs of the village.

Thus in my experience, older men assuming a domineering role was not uncommon, but I must add that the vast majority (if not all) of my interviews with older men in the village were very pleasant and insightful experiences. However, this particular instance was somehow different. This interview was in fact with two people: a man (aged 50–60) and his son (aged 20–30). At the outset, the older man talked over the younger and the latter 'shut down' for the remainder of the interview, deferring to his father, who appeared to have put him in his place. The older man then proceeded to discuss the environmental problems in the area but in such a way that appeared to blame just about everyone but himself for these problems. His manner was very assertive; I felt as if he were delivering a lesson or a sermon to us. His ideas about what 'should be done' about these problems seemed to me somewhat out of touch with village realities.

Not only did my negative emotions toward the man accumulate during the interview, but I also had a compelling sense that my research assistant (a local to the area but educated to university level) was experiencing a similarly negative series of emotions toward the man. This was later confirmed. On concluding the interview and walking back to our house, David stated that: 'Yes, he is an arrogant man! He is somewhat a powerful man in this village so he can say these things. But I am not sure about his ideas, they seem crazy!' We laughed about it. Our dislike for him was largely mutual. My sense of unease at my feelings of dislike was quelled for the time being by the sense of sharing these feelings with David.

The uneasy feeling came back to haunt me. I began to wonder *why* I felt bad about the negative emotions I expressed toward the participant. My conclusion: I had broken the rules, the ethical standards by which researchers conducting their line of business in the Global South are supposed to abide. My pre-fieldwork reading had suggested that in order to redress the power imbalances inherent in this situation, one must first of all be aware of one's positionality vis-à-vis the 'other,' and then attempt to redress these imbalances of power through reciprocity, reflexivity and, where possible, through facilitating forms of empowerment. I (perhaps naively) interpreted these research ethics into a personal ethical and moral code of wishing to sympathize with and positively respond to participants, perhaps to feel some form of solidarity with them. I wished to take a stance which aligned with the needs and aspirations of those that I would study.

At the time I dismissed delving deeper into deconstructing this episode, rejecting an analysis of the incident based, conceivably, on my lack of experience. It is only later reflection and further reading which led me to consider how such emotional, personal encounters during interviews may in fact deconstruct the ethical and moral imperatives which are in some cases implicit, but in other cases explicit, within the development literature. My discomfort was partly emotional and personal, as much as it was entangled with relations of power.

Ethics for development interviews: power and positionality?

The contemporary literature on ethical approaches to development fieldwork, particularly concerning the act of interviewing participants, largely eschews considering how emotion and personality impact on the interview in favor of discussing the imbalance of power relations between the researchers and researched.

This concern for the imbalances of power, and the need to redress these imbalances through commitment to those who participate, has, for geographers, a lineage in both feminist and Marxist geography (Valentine 2005). Feminist scholars have highlighted how a researcher's positionality (their gender, sexuality, race, nationality, age, economic status, etc.) may influence the 'knowledge' produced in and through an interview (Scheyvens and Storey 2003, Scott *et al.* 2006, Momsen 2006), a claim which might be understood as part of a broader recognition in the social sciences that scholars are never 'neutral' or 'unbiased' (Moser 2008).

In development research these insights became prominent in the 1990s. England (1994), examining the 'endemic' exploitation in development fieldwork, questioned how the voices of 'others' could be incorporated into writing without colonizing them and reinforcing patterns of domination. At a similar time, post-development writers (such as Escobar 1995) illustrated how development discourses constructed by Western researchers served to legitimate Western experts, undermining the voices of local people. These critiques prompted a crisis of legitimacy for Western researchers, who were forced to reconsider their role in the research process and how they used particular methods, for fear of reinforcing and justifying unequal power relations which are arguably embedded within colonial histories (Scheyvens and Leslie 2000). Various approaches to tackling these problems have emerged, notably some Westerners withdrawing from research altogether, with others adopting a cultural relativist approach which privileges the knowledge and understandings of those from the Global South (Scheyvens and Storey 2003). Both positions have been critiqued, particularly for romanticizing those 'voices' from the Global South, and for justifying an abdication of responsibility for the Western researcher (Sidaway 1992, Scheyvens and Leslie 2000).

Most textbook guides for researchers in development studies have settled on one response, which is to adopt methodologies and ethics which promote reciprocal relationships and facilitate empowerment. These ideals were embodied in the participatory methods movement (see Chambers 1994a; 1994b; 1994c) and in action research, both of which encouraged a sense of partnership with participants (Cloke *et al.* 2000). For example, Brydon (2006: 26), in a chapter on 'ethical practice,' characterizes the relationship between the researcher and the researched by suggesting that 'the emphasis is on collaboration, facilitation and participation'; however, others go further by describing the researcher's stance as one of 'committed involvement, rather than impartial detachment' (Martin 2000, cited in Scheyvens *et al.* 2003a: 182). Several other authors, writing in key development research texts used commonly by research students, follow this same ethos (Laws *et al.* 2003; Leslie and Storey 2003; Harrison 2006; Momsen 2006; Willis 2006). Indeed, textbook research guides for human geographers also echo these sentiments. Valentine (1997), for example, highlights the import-ance of recognizing one's positionality, of being reflexive, and of redressing power imbalances. Scheyvens and Leslie (2000) go so far as to suggest that the interview itself can encourage empowerment among marginalized women by promoting their self-esteem and affirming self-worth.

In summary, these texts advocate three points to the development researcher. First, the key relationship in the interview is one of (unequal) power. Second, by examining positionality we can encourage greater reflexivity and 'reveal' these power relations, as well as contextualize our interpretations (Moser 2008). Third, I would argue these texts assume that the 'good' development researcher will wish to 'overcome' these inequalities through enacting empowering processes with participants with the aim of stimulating social action (Scheyvens *et al.* 2003a) and following 'the moral imperative to do no harm and hopefully to do good' (Momsen 2006: 47). While I do not disagree with this ethos in a general sense, it makes an analysis of my interview experience cited above rather diffi-cult, as these texts assume 'we' (development researchers) somehow align morally and ethically *with* our participants.

My reading of these texts is that they tend to emphasize the 'positive' rela-tionships which *should* be established with participants. Emotional content of interviews, for example, or clashes in personality, tend to be acknowledged only in passing. Willis (2006: 144) admits that 'there is no right way to conduct inter-views.' Others, such as Harrison (2006: 63), accept that 'what is said by those being interviewed may conflict with the researcher's own views,' and goes on to highlight that 'neither expert nor locals are always right, or always wrong.' Val-entine (1997) and Laws *et al.* (2003) are more explicit: in interviews, partici-pants may express homophobic, racist, or sexist views which may grate against the interviewer's (assumed) morals and ethics. However, these authors sit on the fence in terms of providing an analysis of what the researcher should do in such cases; they merely recognize that such scenarios exist. Such accounts *hint* at the emotional and the personal dimensions emerging in interviews, but do not attempt to analyze how the personalities and emotions of the researcher and the researched might, in fact, play a role. Leslie and Storey (2003) frame the emo-tional content of fieldwork in the Global South as 'culture shock,' which sug-gests a need for 'cultural sensitivity' on the part of the researcher, who should be 'non-judgmental.' Momsen (2006) mentions that there may be 'barriers to mutual understanding,' but again it is unclear if these are general (cultural mis-understandings) or personal, individual, or emotional. While many of these texts therefore suggest that 'there is no perfect formula, no absolutely "right" way of doing things' (Brydon 2006: 29), the sentiment sits uneasily alongside the 'correct' ethics implied in discussions of power and positionality.

I deliberately cite (mostly) development research textbooks above because it is likely that these are where many aspiring and inexperienced researchers gain their knowledge of how to conduct methods 'ethically.' While I do not blame these texts for my naivety in entering the field, they appear to lack the critical tools to analyze negative emotional encounters with interviewees. Even more nuanced understandings of power – for example, by recognizing the ways in which interviewees also have power within the interview process (Cloke *et al.* 2000; Scheyvens and Storey 2003; Harrison 2006), or by understanding that 'criticism of "others" ... can be both relevant and helpful' (Scheyvens and Storey 2003: 8) – do little to advance our grasp of the emotional content of interviews. The analysis here is still on understanding power. Writing which has challenged romanticized notions of 'community' in the Global South, such as Guijt and Shah's (1998) *The Myth of Community*, or those who critique the inherent validity of 'local knowledges' (Briggs 2005; Smith 2011), perhaps goes some way to deconstructing the communitarian ethics of participatory development. Yet these insights have had little impact on the contemporary ethics of development research.

My concern is that these ethics, commendable though the ideas are, have become normative in development research. While the development researcher is no longer seen as 'neutral,' they are instead to aspire to something equally unattainable: a person who is morally and ethically good at all times. This extract from the Developing Areas Research Group's Ethical Guidelines provides an example:

> Members of DARG should endeavour to incorporate the following broad principles in their work in and on the developing world: honesty, integrity, sensitivity, equality, reciprocity, reflectivity, morality, contextuality, non-discriminatory, fairness, awareness, openness, altruism, justice, trust, respect, commitment ... DARG members should thus endeavour to use the research process as a means of reducing these inequalities wherever possible and practicable.
>
> (DARG 2009: 1)

Understanding ethical guidelines as a prescriptive code has been critiqued (Kearns *et al.* 1998; Valentine 2005). Given my own experience, I feel that such moral standards embodied in the DARG ethical guidelines are, for an ordinary human being, mostly unattainable, particularly when doing an interview. Perhaps more interestingly, the broad principles they describe are in many ways personal, traits of personality, and entangled also with experience and emotion. None of the principles listed above have absolute, universal standards; they are indeed highly subjective. Yet, disconcertingly, most of the research development texts avoid an analysis of the personal aspect of ethics. This is a significant concern for interviewing as a method, as it is, perhaps more so than other forms of research methods, an intimate and highly personal encounter.

Emotions and personality in fieldwork in the Global South

Despite this apparent lack of analysis of the emotional and personal content of the interview, particularly in key research textbooks, there is a small but emerging interest in exposing the 'emotional' aspects of interviewing and fieldwork in development geography. Meth and McClymont (2009), for example, give a vivid account of the emotions expressed during interviews in a study of men's experiences of violence in South Africa. One incident, in which a respondent discloses their HIV status, exposes both the respondent's emotions, and the mixed emotions of the interviewers. While the interviewee reveals a positive emotional experience (unloading emotional 'baggage'), unfortunately we do not learn much of the interviewer's feelings when faced with these 'painful realities.' Similarly, Meth and Malaza (2003) express both their own and their participants' emotional distress when researching violence with South African women (see also Brooks, Day, and Skinner in this volume on the emotional impact of interviews).

Outside of direct emotional experiences in the interview, others have reflected on emotions experienced while conducting research in the Global South. Molony and Hammett (2007) describe emotional relationships with research assistants, both positive (becoming friends) and negative (confrontations), although they stop short of describing their own feelings in such situations. Scheyvens *et al.* (2003b) and Cupples and Kindon (2003) both discuss sex and sexuality in the field. Scheyvens *et al.* (2003b) argue that there must be more 'openness' about sex and sexuality, but also articulate concerns about sexual attraction to participants, and the discussion progresses little beyond this statement. Cupples and Kindon (2003: 212) do go further, describing a relationship with a local man and how this involvement altered (positively) their emotional encounters with research participants, suggesting that: 'The field is a site in which our personal and professional roles and relationships converge.' Routledge (2002) offers a particularly lively example of feelings of excitement as he breaks into a hotel in Goa. Routledge is helpfully self-critical of his research 'performance,' particularly of the 'uneasy pleasure' he derives from interviewing individuals while masquerading as a tour operator. However, he too falls short of a full discussion of emotion and personality, instead examining the questions that arise over his positionality when, in fact, it is apparent that Routledge's personality has much to do with his performance. Not every white male academic from the Global North is prepared to masquerade as a tour operator in India and crawl under wire fences.

These examples suggest that the emotional impacts interviews can have on participants and on the interviewer are important (Meth with Malaza 2003; Meth and McClymont 2009). While it is unfortunate that Molony and Hammett (2007: 296) conclude that 'emotional considerations clouded rational academic judgments and serve only to distract from the research process,' others acknowledge that 'emotions can clearly shape the research process quite explicitly' (Meth with Malaza 2003). While Moser (2008: 386) suggests that 'there is a silence regarding how we as individual researchers behave and interact with research subjects, who also have a range of social skills and emotional abilities,' I would disagree. The above accounts *do* break the 'silence' of discussing emotional and personal content, if only in particular ways; yet, perhaps as my own experience of not fully explicable negative feelings toward an interviewee reveal, emotions and personality in the interview remain under-examined.

Analyzing negative emotions and personality in interviews

Thus far I have deliberately drawn from research in the Global South. There is, however, an emerging body of literature from the wider field of human geography which seeks to analyze emotion and personality (rather than positionality and power). It is this literature which I feel offers the most promise for pushing toward an understanding of emotion and personality in the context of the interview.

The fact that emotions, and recounts of them in academic writing, appear at the margins of development geography should perhaps not be a surprise. As both Bondi (2005) and Laurier and Parr (2000) argue, emotion has been pushed to the margins of Western thought and practice, positioned as anti-rational since the Enlightenment. Yet there are a number of scholars who have begun to analyze the emotional geographies of research relationships in practice. I have already mentioned Meth and Malaza (2003) and Meth and McClymont (2009) as some of the very few in development geography who do, but equally important are works by Bondi (2005), Laurier and Parr (2000), Widdowfield (2000), and Burman and Chantler (2004), all of whom have contributed to a developing understanding of emotions as relational. Laurier and Parr (2000: 98) offer a useful conceptualization:

> Emotions can be understood as complex manifestations of corporeal and psychological aspects of human beings which are simultaneously felt and performed as relations between self and world. And in this context, interviewing can be ideally conceived as a ritualized, yet intersubjective encounter which reveals something of such relations and how they are spatially constituted.

This suggests that emotions can, and should, be analyzed and understood relationally. Bondi (2005) also conceptualizes emotions as intrinsically relational, interpersonal, and intersubjective, and draws on psychotherapy and emotional geographies to do so. While I cannot claim to be a scholar of either of those fields, I wish to signpost both of these approaches as useful for offering a way in which to deconstruct contemporary development research ethics.

Understandings of emotional experiences from psychotherapy approaches suggest that any encounter (e.g., the interview) is intrinsically transpersonal, inspired relationally and contextually (Bondi 2005). While a research interview is not a psychotherapeutic encounter, it has similarities: a small number of people discuss issues in depth in an 'intersubjective' encounter (Laurier and Parr 2000). I acknowledged and described my emotions earlier in relation to an individual interviewee. However, working with a psychotherapeutic conceptualization, I could draw from my own interpretations something of that 'betweenness' which constituted the relationship.

For example, I might interpret my negative emotions toward the interviewee as interrelated with his own negative emotional state. These negative feelings may have been directed toward me, or may, judging by his responses, have been fixed on the Tanzanian government, or others he considered to blame for environmental problems. This is not to say that *his* negativity directly caused my own; instead there was a shared negativity in the encounter, one perhaps feeding off the other and vice versa. My distinct feeling, during the interview, that David also was experiencing negative emotions toward the interviewee assisted in a kind of co-production of the negative relationship which we forged with the man. If I were to stretch this analysis further, I might add that my relationship with David was not always one of agreement and friendship (we had some minor disagreements), and perhaps both of us, willing to share in an emotional encounter, encouraged this collective negativity in order to generate a stronger mutuality between us, aware as we were of the tensions in our relationship. Observing the younger man being 'put in his place' early in the interview made me frustrated, and conceivably established my judgment of the older interviewee, such that I had (pre-emptively) initiated a negative relationship with the older respondent based on my sympathy and sense of injustice for the younger man.

Of course, I am not a psychotherapist, but this fragmentary examination perhaps explains how negativity may be reinforced and generated both personally and transpersonally (Bondi 2005). This emotional negativity may not necessarily need to be characterized as 'bad,' as contemporary interview ethics might suggest. My frustration led me to ask a fairly impertinent question, yet this question stimulated further discussion on another point (the respondent's frustration at a lack of political representation). The interview did not 'fail' because I began to dislike the interviewee and challenged him. It generated interesting responses, and, as an unintended outcome, a moment of bonding between myself and David. As Meth and Malaza (2003: 156) describe: 'emotion can be regarded as a 'research resource' in that feelings such as anger can indeed inform scholarship.'

However, this analysis does not seem quite enough to explain my post-interview niggling doubts about my actions and feelings: my fear that I had 'broken the rules' of the interview ethics I aspired to. Laurier and Parr (2000) use the Freudian term *unheimlich* to describe how, in proximity to difference, one can feel 'unsettled,' notably when confronted with a threat to what is known about the self or body. If applied to analysis of emotion, rather than of positionality or power relations, this experience of difference, for me, may have been the occurrence of an emotional encounter which was a threat to my own self-positioning as the 'ethical' researcher. By judging the interviewee's ideas 'out of touch,' and by feeling negativity toward what he said and how he said it, I encountered an emotional state which was 'other' to the expectations I had of myself as the 'ethical interviewer.'

This 'unsettlement' from ethical standards, which arguably encourage the interviewer to become ethically super-human, may not necessarily be a bad thing. An engagement with the unsettling nature of emotional encounters with those in the Global South may allow the interviewer to become *less* than the idealized 'empowerment facilitator': instead, they are a human being who forms emotional relationships (however fleeting) with interviewees. Like Burman and Chantler (2004), I am not an 'emotional expert,' so make no claims that my analysis here is in fact fitting with wider conceptualizations of emotion. What I am hoping to offer is a way of conceptualizing emotion such that it is not 'abnormal' to the process of interviewing in the context of the Global South. As Laurier and Parr (2000) argue, suppression of feeling may ironically end up repressing the experience of socially diverse emotions and the normative rules that produce emotional performances. As such, the experience and attempt at analysis of emotion during interviewing in the Global South may offer insight not only about the interviewer, but also about the interviewees and the context in which they live.

However, I remain uncertain as to whether transpersonal emotional accounts are enough on their own. Much of my negativity was worryingly inexplicable, as I had encounters with similar individuals which did not feel so emotionally charged. The emotional analysis thus far may, unfortunately, read like a 'rationalization' of my emotional negativity in the interview. It is here that, as Moser (2008) suggests, emotional analysis and personality go hand-in-hand. Moser argues that personality may, in some cases, be more important that positionality in the interview. She suggests that her own sociability and extrovert personality were essential for her fieldwork in Indonesia, while she witnessed other researchers, whose personalities were more introverted, struggling despite the fact that they had more 'in common' with participants based on their ethnic and cultural background.

Although I had many successful interviews in which I was able to build a sense of emotional rapport with interviewees, I was clearly not capable of performing this role at all times. Indeed, it is apparent from the extracts of the interview that there was an 'edginess' to my personality which surfaced during its course. I disagree with Moser's (2008: 386) conclusion that 'the solution here is not to attempt to change one's personality to fit a fieldwork situation but to engage in fieldwork which utilizes one's strengths.' This is unhelpful for those with little experience, and suggests that certain emotional intelligences cannot be 'learnt.' The incident I describe exposed some of the more negative aspects of my personality, yet this was only one performance of personality (see also Bryant in this volume on suppressing personality during ethnographic fieldwork). Indeed, in many (if not most) other interviews, I drew, I hope, on a more empathic, personable side of my character.

Drawing from studies in psychology, Moser further illustrates that integrated approaches to personality situate the individual within social, cultural and historical contexts, but at the same time these approaches do not reject the possibility that universal, cross-cultural truths exist about human personality. Indeed, some personality traits may in fact vary little across cultures (Moser 2008). I would argue that the focus of interview ethics on positionality and power may serve to entrench the differences between the interviewer and participant. Attending to personality serves instead to emphasize some of the commonalities of human experience and emotion. Just as positive aspects of personality may be of significance for the success of research (Meth and McClymont 2009), so may the negative aspects of human personality be equally important points of commonality and departure between individual human beings, and therefore of significance for analysis in the context of an interview encounter. Indeed, the shared experience of negativity between myself and my research assistant toward our interviewee perhaps says something about a common facet of human personality: the ability to find the ideas of 'others' both arrogant and preposterous, despite the fact that these 'others' may be significantly disadvantaged and marginalized when compared to oneself. I am not arguing that this is 'good,' but I am accepting that such a feeling is possible within my own personality and emotional range.

Toward new interview ethics?

I have attempted to draw on my own experience of negative emotions during an interview with a respondent in Tanzania to highlight what I see as some of the key problems with existing interview ethics. My analysis is necessarily fragmentary and incomplete; I do not make claims to be a scholar of emotions or personality, but I do wish to highlight both fields as offering potential for allowing current and future researchers to move beyond thinking of interview ethics in the Global South as being just about positionality and power.

In some respects I am writing about this experience for new researchers who may be, as I was, naive about what they may encounter in the intimate, intersubjective world of the interview. My naivety was founded partially on a reading of contemporary ethical standards, which I feel, in their present form, set an unattainable standard for me as an inexperienced PhD researcher. The revelation for me may only be that disadvantaged, poor, and marginalized communities in the Global South are comprised entirely of human beings with whom researchers will have human, emotional relations. This may sound obvious, but working from the ethics of many contemporary research textbooks left me with the impression that encounters with the people I wanted to work with would tend my emotions toward a willingness to sympathize, reciprocate, and commit to forms of solidarity with those I interviewed, to work toward the lofty goal of facilitating empowerment, to redress the inequalities of our relative positions of power.

Yet I often found myself relating to individuals in a different way, through relational encounters suffused with emotion, and which drew on my personality as much as their own. My prescribed positionality and relative power-position were superseded by my personality and emotions. It is only from reading beyond the consensus on research ethics found commonly in development research text-books that I have begun to find an adequate way to conceptualize and analyze the significance of these emotional states. Indeed, if 'feeling and thinking are two sides of the same coin' (Bondi 2005: 444), then we should be thinking of the interview as an emotional space as much as it is a 'thinking space.'

As Moser (2008) and Burman and Chantler (2004) advise, our understanding of personalities and emotional abilities will always be fragmented, partial, impressionistic, and anecdotal, but this does not mean they are 'unknowable' nor, I would add, therefore less relevant than considerations of power and positionality. Yet 'knowing' these emotions in the development interview requires a deconstruction of the gold standard of development research ethics. Doing away with prescriptive codes of ethics is part of the battle here, particularly those which necessitate 'for academics to be *with* resisting others as well as *for* them' (Routledge 2002: 478). While such commitments may be appropriate at a very broad level, such as dedication toward decreasing inequalities, or being person-ally relevant *for* particular political, ethical, or moral causes, such prescriptive assumptions about the nature of all researchers working in the Global South break down at the individual, interpersonal level. Assuming that one will inher-ently feel sympathy, commitment, trust, and respect (DARG 2009) toward an individual purely because of their positionality and situation of relative power denies the fact that emotional and interpersonal relations which will emerge during the interview may inspire quite opposite feelings. Such ethics are, in a sense, de-humanizing, positing individuals as positionalities rather than as indi-vidual human beings.

While this kind of very personal discussion can tend toward narcissism, Meth and Malaza (2003) insist that an important aspect of discussing emotions is to provide support for other researchers, perhaps just as important as providing a further aspect of analysis for research results. Public discussions of the emo-tional and personal attributes of interviewing will also further reduce the poten-tial for honesty to be read by others as vulnerability and poor research performance (Cloke *et al.* 2000), and instead begin to conceptualize emotion and personality as a normal part of the research encounter. Sharing emotional research experiences in a meaningful manner is perhaps the first step toward a relational, personal ethic of doing research in the Global South.

Recommended reading

Bondi, L. (2005) 'Making connections and thinking through emotions: Between geography and psychotherapy,' *Transactions of the Institute of British Geo-graphers*, NS 30: 433–448.

Although this paper is not focused on research in the Global South, it offers considerable insight into how to conceptualize the emotional and personal content of interviews and other research encounters.

Cloke, P., Cooke, P., Cursons, J., Milbourne, P. and Widdowfield, R. (2000) 'Ethics, reflexivity and research: Encounters with homeless people,' *Ethics, Place and Environment*, 3 (2): 133–154.

Again not in a 'Global South' setting, but invaluable for the honest approach the authors take toward writing about the emotions they experienced when conducting interviews with homeless people.

Meth, P. and McClymont, K. (2009) 'Researching men: The politics and possibilities of a qualitative mixed-method approach,' *Social and Cultural Geography*, 10 (8): 909–925.

Highly recommended for the accounts of interviewing and conducting mixed method research on men's experience of violence in South Africa. The paper is revealing of the complex emotional encounters which researchers might face when interviewing, particularly on sensitive topics.

Moser, S. (2008) 'Personality: A new positionality?,' *Area*, 40 (3): 383–392.

While I do not agree with some of the conclusions of this piece, the discussion of personality and positionality is excellent, and provides a solid foundation from which to begin analyzing how one's personality may impact on research in the Global South.

References

Bondi, L. (2005) 'Making connections and thinking through emotions: Between geography and psychotherapy,' *Transactions of the Institute of British Geographers*, NS 30: 433–448.

Briggs, J. (2005) 'The use of indigenous knowledge in development: Problems and challenges,' *Progress in Development Studies*, 5 (2): 99–114.

Brydon, L. (2006) 'Ethical practice in doing development research,' in V. Desai and R. B. Potter (eds), *Doing development research*, London: Sage.

Burman, E. and Chantler, K. (2004) 'There's no-place like home: Emotional geographies of researching "race" and refuge provision in Britain,' *Gender, Place & Culture: A Journal of Feminist Geography*, 11 (3): 375–397.

Chambers, R. (1994a) 'The origins and practice of participatory rural appraisal,' *World Development*, 22 (7): 953–969.

Chambers, R. (1994b) 'Participatory rural appraisal (PRA): Analysis of experience,' *World Development*, 22 (9): 1253–1268.

Chambers, R. (1994c) 'Participatory rural appraisal (PRA): Challenges, potentials and paradigm,' *World Development*, 22 (10): 1437–1454.

Cupples, J. and Kindon, S. (2003) 'Far from being "home alone": The dynamics of accompanied fieldwork,' *Singapore Journal of Tropical Geography*, 24 (2): 211–228.

DARG (Developing Areas Research Group) (2009) *DARG ethical guidelines*, available at: www.devgeorg.org.uk/?page_id=799 (accessed 14 January 2013).

England, K. (1994) 'Getting personal: Reflexivity, positionality and feminist research,' *The Professional Geographer*, 46: 80–89.

Escobar, A. (1995) *Encountering development: The making and unmaking of the third world*, Princeton, NJ: Princeton University Press.

Guijt, I. and Shah, M. K. (eds) (1998) *The myth of community: Gender issues in participatory development*, London: ITDG Publishing.

Harrison, M. E. (2006) 'Collecting sensitive and contentious information,' in V. Desai and R. B. Potter (eds), *Doing development research*, London: Sage.

Kearns, R., Le Heron, R. and Romaniuk, A. (1998) 'Interactive ethics: Developing understanding of the social relations of research,' *Journal of Geography in Higher Education*, 22: 297–310.

Laurier, E. and Parr, H. (2000) 'Disability, geography and ethics,' *Philosophy and Geography*, 3 (1): 98–102.

Laws, S., Harper, C. and Marcus, R. (2003) *Research for development: A practical guide*, London: Sage.

Leslie, H. and Storey, D. (2003) 'Entering the field,' in R. Scheyvens and D. Storey (eds), *Development fieldwork: A practical guide*, London: Sage.

Meth, P. and McClymont, K. (2009) 'Researching men: The politics and possibilities of a qualitative mixed-method approach,' *Social and Cultural Geography*, 10 (8): 909–925.

Meth, P. with Malaza, K. (2003) 'Violent research: The ethics and emotions of doing research with women in South Africa,' *Ethics, Place and Environment*, 6 (2): 143–159.

Molony, T. and Hammett, D. (2007) 'The friendly financier: Talking money with the silent assistant,' *Human Organisation*, 66 (3): 292–300.

Momsen, J. H. (2006) 'Women, men and fieldwork: Gender relations and power structures,' in V. Desai and R. B. Potter (eds), *Doing development research*, London: Sage.

Moser, S. (2008) 'Personality: A new positionality?,' *Area*, 40 (3): 383–392.

Routledge, P. (2002) 'Travelling east as Walter Kurtz: Identity, performance and collaboration in Goa, India,' *Environment and Planning D: Society and Space*, 20: 477–498.

Scheyvens, R. and Leslie, H. (2000) 'Gender, ethics and empowerment: Dilemmas of development fieldwork,' *Women's Studies International Forum*, 23 (1): 119–130.

Scheyvens, R. and Storey, D. (2003) 'Introduction,' in R. Scheyvens and D. Storey (eds), *Development fieldwork: A practical guide*, London: Sage.

Scheyvens, R., Novak, N. and Scheyvens, H. (2003a) 'Ethical issues,' in R. Scheyvens and D. Storey (eds), *Development fieldwork: A practical guide*, London: Sage.

Scheyvens, R., Scheyvens, H. and Murray, W. E. (2003b) 'Working with marginalised, vulnerable or privileged groups,' in R. Scheyvens and D. Storey (eds), *Development fieldwork: A practical guide*, London: Sage.

Scott, S., Miller, F. and Lloyd, K. (2006) 'Doing fieldwork in development geography: Research culture and research spaces in Vietnam,' *Geographical Research*, 44 (1): 28–40.

Sidaway, J. D. (1992) 'In other worlds: On the politics of research by "First World" geographers in the "Third World",' *Area*, 24: 403–408.

Smith, T. A. (2011) 'Local knowledge in development (geography),' *Geography Compass*, 5 (8): 595–609.

Valentine, G. (1997) 'Tell me about …: using interviews as a research method,' in R. Flowerdew and D. Martin (eds), *Methods in human geography: A guide for students doing a research project*, Harlow: Prentice Hall.

Valentine, G. (2005) 'Geography and ethics: moral geographies? Ethical commitment in research and teaching,' *Progress in human geography*, 29: 483–487.

Widdowfield, R. (2000) 'The place of emotions in academic research,' *Area*, 32: 199–208.

Willis, K. (2006) 'Interviewing,' in V. Desai and R. B. Potter (eds), *Doing development research*, London: Sage.

14 Whose knowledge, whose benefit? Ethical challenges of participatory mapping

Experiences from fieldwork on mapping community values on land in Zanzibar

Nora Fagerholm

> See, this is how we use the land! The fields are all over and we lose the forests if we cultivate like this.

This extract from my fieldwork diary was a worried farmer's interpretation of an aerial image map that showed the results of a participatory mapping exercise. His statement captures some of the ethical dilemmas that I faced when conducting my doctoral fieldwork in Zanzibar. My specific interest was participatory mapping, i.e., using local communities to create spatial representations of their land and resource use as well as the diversity of perceptions and values attached to their everyday landscapes (Tuan 1977; Zube 1987; Williams and Patterson 1996; Stephenson 2008). The participatory approach to mapping, which has been popular in development studies since the 1960s, is designed in part to counter some of the power relations inherent in any mapping process. Since the development of cartography as a science, maps have been associated with power and politics (Wood 2010). Getting local people to create spatial representations of what is important to them rather than what outsiders think is important to them is a challenge to the traditional approach to cartography and has been particularly important in development research and practice across the Global South (Mikkelsen 2005; Chambers 2008).

My research combined community participation with the use of geospatial technologies – known as participatory geographical information systems (PGIS) – which enabled the data collected from stakeholders to be stored, analyzed, and displayed in digital format (Rambaldi *et al.* 2006b; Sieber 2006; Dunn 2007). Although PGIS is an exciting methodology that brings together qualitative and quantitative data, it is laden with ethical dilemmas (McCall 2003). While the process is intended to be participatory and empowering by involving local people, processing the data within GIS systems necessarily requires technical expertise and the data may be represented in formats that are alien to those who originally provided it. Thus there are various challenges of ensuring that the process leads to sustainable gains rather than losses for the participants by ensuring that the process promotes collaboration and empowerment (Craig *et al.* 2002; Chambers 2006). In this chapter, I explore some of the ethical dilemmas that I faced when conducting participatory GIS in Zanzibar, Tanzania.

My research focused on two rural case study locations in Zanzibar where I combined semi-structured interviews with participatory mapping on aerial images. The aim was to capture local knowledge on the uses, values, and perceptions of land and forest resources in a spatial context. I then undertook descriptive, statistical, and spatial analysis of the data which enabled me to collate individual interpretations of the landscape onto map representations showing multiple perceptions of the same landscape. In my view, this PGIS methodology is a really effective way to capture the multidimensional practices across a particular landscape and to compare individual with collective landscape values.

While a participatory approach to research is seen as being particularly appropriate in the context of sustainable landscape development (e.g., Luz 2000; Raquez and Lambin 2006; Selman 2006), it is embedded with ethical dilemmas. While I found the codes of ethics in anthropology, ethnobiology, and GIS (American Anthropological Association 2009; International Society of Ethnobiology 2006; URISA Board of Directors 2003) very helpful, the reality is that every research setting is unique and, in the end, the researcher must subjectively judge what good practice is and make certain ethical choices. For me, the most helpful guide toward good practice was that in Rambaldi *et al.* (2006a), which is reproduced in Figure 14.1. The authors compiled a series of 'who?' and 'whose?' questions which are broadly divided into the before (planning), during (data collection), and after (results) stages of research. Although their guide was developed with reference to PGIS, I feel that it is relevant across a range of methodologies dealing with participation and local knowledge. In the next sections I will explore some of the 'who?' and 'whose?' questions in the context of my own field experiences and discuss how I negotiated them.

Whose problems, questions, and perspectives?

I was part of a multicultural research team consisting of Finnish academics, Tanzanian academics, and Zanzibarian forest officers all working as researchers and facilitators. My university department has been doing landscape research in Zanzibar in collaboration with the Department of Forestry and Non-Renewable Natural Resources (DFNR) in Zanzibar and the Department of Geography at the University of Dar es Salaam in Tanzania for more than ten years, and there have been several research projects. From the beginning my doctoral research was related to these activities, and during the latter part it was funded by the Zanzibar research project. Thus I designed my research project – including the motivation, aims, case studies, and data collection – under the umbrella of the wider interests of the team.

I selected two rural areas for the study localities. The differences in biophysical and geographical conditions and forest management policy implementation formed the basis for the diverse land uses between these areas and contributed to my selection of them from among the six study areas where the University of Turku Zanzibar research team has been conducting research over recent years. The areas were originally suggested as research sites by the Zanzibar forest administration because they were communities facing different kinds of

STAGE 1: PLANNING

Who participates?
Who decides on who should participate?
Who participates in whose mapping?
    ... and who is left out?

Who identifies the problem?
Whose problems?
Whose questions?
Whose perspective?
... and whose problems, questions,
    perspectives are left out?

STAGE 2: THE MAPPING PROCESS

**Whose voice counts? Who controls
the process?**
Who decides what is important?
Who decides, and who should decide, on
    what to visualise and make public?
Who has visual and tactile access?
Who controls the use of information?
And who is marginalised?

Whose reality? And who understands?
Whose reality is expressed?
Whose knowledge, categories,
    perceptions?
Whose truth and logic?
Whose sense of space and boundary
    conception (if any)?
Whose (visual) spatial language?
Whose map legend?
Who is informed what is on the map?
    (Transparency)
Who understands the physical output?
    And who does not?
And whose reality is left out?

**STAGE 3: RESULTING INFORMATION
CONTROL, DISCLOSURE AND
DISPOSAL**

Who owns the output?
Who owns the map(s)?
Who owns the resulting data?
What is left with those who generated the
    information and shared their
    knowledge?
Who keeps the physical output and
    organises its regular updating?

Whose analysis and use?
Who analyses the spatial information
    collated?
Who has access to the information and
    why?
Who will use it and for what?
And who cannot access and use them?

ULTIMATELY ...

**What has changed? Who benefits from
the changes? At whose costs?**
Who gains and who loses?

**Who is empowered and who is
disempowered?**

Figure 14.1 Compilation of 'who?' and 'whose?' questions as a pathway leading toward
    a good participatory mapping (PGIS) practice (source: Adapted from
    Rambaldi *et al.* 2006a: 108, Box 1).

challenges in the management of land and resources. Once I had decided my
case study locations, I organized community meetings where I was able to
identify the main concerns relating to the management of natural resources in
these subsistence-based communities. This interaction with the communities fed
into the planning for the fieldwork.

Although the preparatory stage of selecting field locations was partly led by a local governmental organization, I was responsible for ensuring that the theoretical framework and research questions were academically rigorous and would stand up under the scrutiny of my supervisors and examiners. I also had to develop the typologies capturing the diversity of the subjective everyday landscape practices and experiences that I would use for the mapping process. While existing typologies found in the literature were very helpful (e.g., Millennium Ecosystem Assessment 2003; De Groot *et al.* 2009), every culture and context is different, so I liaised with local members of the research team and considered other studies (e.g., Brown 2005; Tyrväinen *et al.* 2007) to try to ensure that our typology would be appropriate in the case study localities.

The chosen typologies comprised four social landscape values (subsistence, aesthetic, traditional, leisure) and 11 landscape services which were divided into the material (e.g., food; raw materials; geological resources; and fuel, medicinal and ornamental resources) and non-material (aesthetics; social relations; and spiritual, religious, intrinsic, cultural and intrinsic values). Although these typologies were informed by the experience of other researchers and the advice of local people familiar with the research location, they were ultimately my responsibility. There will always be a risk that a typology developed by a Western person – and particularly one concerning perceptions and values – may omit some essential aspects that one is not aware of. However, issues that emerged during the interviews appeared to be minor and I was confident about my chosen approach for typology formulation. An alternative approach would have been, for example, to use a grounded perspective method through which the community defines the values (e.g., Stephenson 2008).

Designing academically rigorous research within a theoretical framework and planning participatory mapping activities in the field is simultaneously a top-down and bottom-up approach. Hence it raises a variety of ethical questions about whose problems, questions, or perspectives are being considered in the research.

Who controls the process and whose voice counts?

Participatory research, by its very nature, involves people but not everyone. Thus an important ethical issue to consider at the planning stage is who is included and who excluded in the process. Within our two case study areas I wanted to include each sub-village and, within each of these, a sample of participants. Unfortunately, census data was not available to ensure geographically-balanced random sampling; instead I calculated the number of participants representing each sub-village relative to the number of inhabitants or buildings there. The village leaders were then asked to select a specific number of participants; however I specified the gender balance and age structure. So the selection of participants was based on a combination of estimated statistics, deliberate targeting, and personal selection.

While the geographical representativeness of the sample (which is an especially important characteristic for GIS analysis) would stand up to academic scrutiny, the other stages of the filtering process have various shortcomings. These raise a number of issues about who actually controls the process of selection and whether particular participants were chosen for political reasons or on the basis of being perceived as being more interesting. For example, village leaders were instructed to select only one person in each household to be interviewed. It is a traditional clan-based society (formed around the same family or a couple of families) and so it is quite possible that individuals from different households were close relatives of the village leader. However, I consider it unlikely that this bias had too much influence on the validity of the sample.

Participation was supposed to be on a voluntary basis, but I cannot be certain whether some of the selected participants had been persuaded to take part. Although I was following ethical guidelines for informed consent (e.g., International Society of Ethnobiology 2006), I had placed the decision-making in the hands of the village leaders because of the absence of census data which would have enabled random sampling. Additionally, participants received a small monetary compensation after their interviews, and I cannot be sure whether this influenced the situation in terms of motivating their participation or prejudicing their responses (Mikkelsen 2005: 344).

Mapping landscape practices and experiences is very subjective, so I preferred to conduct single-informant interviews rather than group mapping exercises. The latter can have their advantages, but the main risk is that some participants dominate the process leading to an imbalance in voices heard (Chambers 2002; Mikkelsen 2005; Bernard *et al.* 2011). In retrospect, I consider this choice of individual interviews to be crucial when aiming to derive collective characteristics and spatial patterns of social landscape values and landscape services based on their individually identified meanings.

In addition, interpersonal power relations affect the interview process and are linked to ethical issues (Madge 1997): for example, what informants are willing to share depends on the trust they place in the researcher. Thus the informant ultimately has the power to decide what kind of a voice is heard. In my case, this was especially relevant when considering that the interviews were partly facilitated by the local forest officers in the research team. Although these forest officers were engaged with forest planning and management in Zanzibar, participants were told that the involvement of the forest officers in the interviews was purely in a research assistant capacity and that the participants' responses, particularly concerning the use of resources within conservation areas, would not be judged or used in other contexts by the forest officers in their official capacity. In fact, during the interview campaigns, I had the feeling that the communities were indeed seeing us as a research team and not specifically questioning the role of the forest officers.

Overall, some of these ethical challenges faced during sampling and interviewing raise critical questions about who controls the process and whose voices are heard. In my experience, no matter how you try to follow textbook guidance, the reality in the field is less straightforward. There were possible biases in my sampling dictated by local circumstances and in the interview responses due to power relations, but I have to accept that fieldwork will never be a perfect academic exercise and the best we can do is to be aware of the ethical issues.

Who will use the data and for what purpose?

Although my research involved gathering information from community members, once this knowledge was collated and represented on a map with a legend, it became public knowledge and therefore open to use (or abuse) by outsiders. Ironically this had the potential to undermine the very goals of participatory mapping. For example, maps of local knowledge on land could potentially weaken the existing common property management systems, or the data might be used for extracting more taxes or for exclusive land privatization (Abbott *et al.* 1998; Fox *et al.* 2005).

One particular ethical challenge that I faced involved considering how to represent individual views and local knowledge in a way that would cause no harm. It is a question of carefully balancing geographical precision and accurate academic analysis with a certain level of ambiguity and abstraction in order to protect the participants and their data (McCall 2006). While researchers writing about interviewees can use pseudonyms, the task of obscuring individual responses on maps is more difficult. The data was collected on local-scale aerial image maps (1:5,000/1:12,000) through polygon delineation and point placement methods. My analysis relied mostly on distance-based approaches and on the spatial aggregation of polygons/points to identify areas of high intensity and clustering (Brown 2005; Brown and Pullar 2011). I was not interested in exact locations but rather in the spatial landscape patterns, so I did not need to represent the data with such a precision that it could be connected to specific places or individual informants. Instead, I applied the resolution of at least 50- and 200-meter grid cells which were intended to maintain the confidentiality of the informants yet to be precise enough to represent spatial patterns. For example, the distribution of individual sacred places – some of which were well-known sites with long traditions but others of which were used by only a few people – is indicated, but simultaneously the private nature of these traditional sites is respected.

These challenges of analyzing data collected from individual people during participatory mapping and representing multiple and collective perspectives in spatial form raised a variety of ethical questions about who uses the data and for what purpose. The particular challenge for me was to ensure that I did not cause the individuals or communities any harm, either between themselves or with outsiders, by revealing information which could be appropriated and misused.

Who owns the data?

The collection of data through participatory mapping and its analysis was not the end of the process. In line with good ethical practice in research, and perhaps even more so in participatory research, I wanted to share the results with the participants (International Society of Ethnobiology 2006).

I organized community meetings and focus-group discussions for those informants that had been interviewed in each case study location. These meetings were not only for the purpose of showing the maps that I had generated from my analysis but also an opportunity to discuss critical issues regarding the use of natural resources, levels of collaboration, and gender differences in resource management. The discussions were also useful for my research because they allowed me to deepen my interpretation of the results. More importantly for ethical research, however, the discussions were a means to empower communities to have a wider understanding of their land and resources beyond their immediate households.

After a quiet start, most meetings evolved into lively discussions. The participants were very enthusiastic about the maps, on which they could look at the distribution of the landscape values and services overlaid with the aerial image and interpret the spatial patterns. I explained that for confidentiality reasons, the maps did not show any individual inputs. The participants were happy with the general results but, not surprisingly, seemed interested in looking at them from their home perspective. As noticed already during the interviews, the community stakeholders were able to identify places and areas on the aerial image maps with little support; these maps engaged even illiterate informants (see also Taylor *et al.* 2006; Bernard *et al.* 2011), in contrast to abstract map representations (Zurayk *et al.* 2001). Personal differences were evident, though, and on some occasions facilitators helped informants who had less cartographical literacy.

Of course these discussions were not without ethical challenges. The native Swahili speakers – the local forest officers and the Tanzanian academics – who were facilitating the discussions had to be aware of various sensitivities surrounding land management issues and avoid inflaming any potential conflicts among the participants. These included, for example, corruption related to the sales of land plots where the village leaders often make profit, and the competition between communities and recently established international hotels over the use of coastal land and resources. Within our research team, we prepared ourselves by discussing issues of potential conflict before the community interaction meetings.

Rather than the community meetings being a single event for 'giving back' to participants, the maps and printouts with Swahili information and legends used in the meetings were given to the communities. On some occasions people also asked for them. The hope was that the village administration or village conservation committees could store the materials and use them in the future when discussing the use and valuation of forest resources. Of course, I cannot be sure if this gesture, which aimed to strengthen the ownership of the local knowledge, has genuinely made the data publicly available in the communities. Copies of the maps were also left with the forest administration in Zanzibar to exemplify the potential use of place-based local knowledge in community forest management.

By following ethical guidelines and using reflective meetings and printed resources to 'give back' the local knowledge to those who had generated it, I

tried to address the question of who owns the data. This was not without challenges, as some of the data was potentially political and divisive. We can never
truly know the impact and outcomes of our research; we can only hope that we
have done our best (see also Staddon in this volume for further discussion on
'giving back').

Who is empowered?

In development research and practice, the term 'participation' usually goes hand-
in-hand with 'empowerment' and 'capacity-building' (Kesby 2000; Reed 2008).
In the case of my research I hope that the participatory mapping exercises did
serve to empower communities and build capacity at different levels and in
several ways.

In terms of empowerment, participants without experience in land management contributed valuable data about practices and values attached to their land
and living space. Although the sampling procedure had flaws (as discussed
earlier) we did receive positive feedback from the village leaders, who appreciated that participants represented the community as a whole and not only those
who were regularly engaged in environmental issues at the local level. However,
not everyone could participate so not everyone was automatically empowered.
The use of aerial image maps on which community stakeholders could identify
places with little support proved to be very successful in participatory mapping
and data reflection. Furthermore the mapping exercise generated data in a legitimate and official format that was later integrated with other government and
expert GIS data sets. This means that the communities and their opinions can be
taken seriously on a wider stage. Hence this process of involving community
members in the production of knowledge in a spatial context and valuing their
contributions, particularly from those with limited written literacy, was hopefully empowering.

The project also contributed toward capacity-building at different levels. At
community level the reflective meetings often came around to discussing issues
of the sustainability of the village landscape and natural resources; this forum
provided a place for participants to better understand broader issues and their
role and rights in them. As mentioned earlier, I worked as part of a research
team; by working together on planning, analysis, and reporting we hope to have
enabled the Tanzanian members of the team to develop new skills that will build
their capacity as professional researchers in natural resource management. Furthermore, the local government benefited because they subsequently learned to
apply the participatory mapping method, the concept of place-based local knowledge, and the case study results in community forest management processes in
Zanzibar.

Thus at different levels and in different ways I hope that my work was a
process which empowered the local community and built capacity for longer-
term sustainability in the area.

Some final thoughts on who gains and who loses

Of course the primary aim of my involvement in this work in Zanzibar was to undertake a successful doctoral research project and to make a scientific contribution to the field of applying participatory and GIS techniques for the mapping and spatial analysis of the diversity of subjective everyday landscape practices and experiences (Fagerholm and Käyhkö 2009; Fagerholm *et al.* 2012; Fagerholm *et al.* 2013; Käyhkö *et al.* 2013). Even if we researchers are motivated by additional – and perhaps more altruistic – reasons, we must be open and honest about the fact that we stand to gain the most from our research in terms of a qualification, some publications, and hopefully a step up the career ladder. In the light of this, I did explain the purpose of the research and the potential lack of concrete benefits to the participants in order not to raise false expectations.

However, this is not to say that while I have gained the local communities and individual participants have necessarily lost. I hope that there have been some positive outcomes for them too. The key to successful PGIS practice is identified as effective participation, which should be seen as a carefully planned and demand-driven process (Rambaldi *et al.* 2006b; Reed 2008; Corbett and Rambaldi 2009). As outlined in the previous section, the empowerment and capacity-building of the community, the research team, and the local government were all built into the project. The case studies were small stepping-stones toward developing land and resource management policies that identify the value of stakeholder data in spatial form and that institutionalize place-based local knowledge in planning. This will enhance the long-term viability of natural resources in Zanzibar and is equally applicable to other tropical areas.

As highlighted at the beginning, all mapping is inherently related to power. No matter how well-intentioned I was in my participatory approach, my involvement was not politically neutral and there may be some unintended consequences of the research activities that I will never know (Madge 1997; Corbett and Rambaldi 2009). Thus the question of who gains and who loses may never be fully answered. I have grown familiar with the Zanzibar context and come closer to the insider's perspective. Yet I acknowledge being an outsider and having a visitor's gaze on a landscape already laden with particular cultural values, attitudes, ideologies, and expectations (Wylie 2007). Although my methodological focus was participatory GIS, many of the ethical dilemmas about representation, voices, giving back, empowerment, and so on that I faced are common to other qualitative and participatory methodologies and reappear in the field experiences of others relayed throughout this volume.

Recommended reading

Chambers, R. (2006) 'Participatory mapping and geographic information systems: Whose map? Who is empowered and who disempowered? Who gains and who loses?,' *The Electronic Journal of Information Systems in Developing Countries*, 25: 1–11.

Discusses the evolution of the integration of participatory methods with spatial technologies and raises concern over skilled facilitation and the importance of practical ethics.

McCall, M. (2003) 'Seeking good governance in participatory-GIS: A review of processes and governance dimensions in applying GIS to participatory spatial planning,' *Habitat International*, 27: 549–573.

This is a paper addressing the issues of ownership, legitimacy, and local knowledge when applying GIS to participatory planning and for community empowerment.

Rambaldi, G., Chambers, R., McCall, M. and Fox, J. (2006) 'Practical ethics for PGIS practitioners, facilitators, technology intermediaries and researchers,' *Participatory Learning and Action*, 54: 106–113.

This is a paper guiding toward good practice and appropriate ethical choices for those practicing or wanting to practice participatory mapping, especially PGIS.

References

Abott, J., Chambers, R., Dunn, C., Harris, T., de Merode, E., Porter, G., Townsend, J. and Weiner, D. (1998) 'Participatory GIS: Opportunity or oxymoron?,' *PLA Notes*, 33: 27–33.
American Anthropological Association (2009) *Code of Ethics (approved February 2009)*, available at: www.aaanet.org/cmtes/ethics/Ethics-Resources.cfm (accessed 22 November 2012).
Bernard, E., Barbosa, L. and Carvalho, R. (2011) 'Participatory GIS in a sustainable use reserve in Brazilian Amazonia,' *Applied Geography*, 31: 564–572.
Brown, G. (2005) 'Mapping spatial attributes in survey research for natural research management: Methods and applications,' *Society and Natural Resources*, 18: 17–39.
Brown, G. and Pullar, D. (2011) 'An evaluation of the use of points versus polygons in Public Participation Geographic Information Systems using quasi-experimental design and Monte Carlo simulation,' *International Journal of Geographical Information Science*, 1: 1–16.
Chambers, R. (2002) *Participatory workshops: A sourcebook of 21 sets of ideas and activities*, London: Earthscan.
Chambers, R. (2006) 'Participatory mapping and geographic information systems: Whose map? Who is empowered and who disempowered? Who gains and who loses?,' *The Electronic Journal of Information Systems in Developing Countries*, 25: 1–11.
Chambers, R. (2008) *Revolutions in development inquiry*, London: Earthscan.
Corbett J. and Rambaldi, G. (2009) 'Geographic information technologies, local knowledge, and change,' in M. Cope and S. Elwood (eds), *Qualitative GIS: A mixed methods approach*, Sage, London.
Craig, W. J., Harris, T. M. and Weiner, D. (eds) (2002) *Community participation and geographic information systems*, London: Taylor and Francis.
De Groot, R. S., Alkemade, R., Braat, L., Hein, L. and Willemen, L. (2009) 'Challenges in integrating the concept of ecosystem services and values in landscape planning,' *Ecological Complexity*, 7: 260–272.

Dunn, C. E. (2007) 'Participatory GIS: A people's GIS?,' *Progress in Human Geography*, 31: 616–637.

Fagerholm, N. and Käyhkö, N. (2009) 'Participatory mapping and geographical patterns of the social landscape values of rural communities in coastal Zanzibar, Tanzania,' *Fennia*, 187: 43–60.

Fagerholm, N., Käyhkö, N., Ndumbaro, F. and Khamis, M. (2012) 'Community stakeholders' knowledge in landscape assessments: Mapping indicators for landscape services,' *Ecological Indicators*, 18: 421–433.

Fagerholm, N., Käyhkö, N. and Van Eetvelde, V. (2013) 'Local level landscape characterization as a medium of participatory community forest management: Integrating and representing expert and stakeholder knowledge,' *Environmental Management*, DOI 10.1007/s00267–013–0121-x.

Fox, J., Suryanata, K. and Hershock, P. (2005) *Mapping communities: Ethics, values, practice*, Honolulu: East-West Center.

International Society of Ethnobiology (2006) *ISE Code of Ethics (with 2008 additions)*, available at: www.ethnobiology.net/ethics.php (accessed 22 November 2012).

Käyhkö, N., Fagerholm, N. and Mzee, A. J. (2013) 'Local farmers' place-based forest benefits and government interventions behind land cover and forest transitions in Zanzibar, Tanzania,' manuscript submitted to *Journal of Land Use Science*, DOI 10.1080/1747423X.2013.858784.

Kesby, M. (2000) 'Participatory diagramming: Deploying qualitative methods through an action research epistemology,' *Area*, 32: 423–435.

Luz, F. (2000) 'Participatory landscape ecology: A basis for acceptance and implementation,' *Landscape and Urban Planning*, 50: 157–166.

Madge, C. (1997) 'The ethics of research in the "Third World",' in E. Robson and K. Willis (eds), *Postgraduate fieldworks in developing areas: A rough guide*, Developing Areas Research Group Monograph No. 9.

McCall, M. (2003) 'Seeking good governance in participatory-GIS: A review of processes and governance dimensions in applying GIS to participatory spatial planning,' *Habitat International*, 27: 549–573.

McCall, M. (2006) 'Precision for whom? Mapping ambiguity and certainty in (participatory) GIS,' *Participatory Learning and Action*, 54: 114–119.

Mikkelsen, B. (2005) *Methods for development work and research: A new guide for practitioners*, New Delhi: Sage Publications.

Millennium Ecosystem Assessment (2003) *Ecosystems and human well-being: A framework for assessment*, Washington, DC: Island Press.

Rambaldi, G., Chambers, R., McCall, M. and Fox, J. (2006a) 'Practical ethics for PGIS practitioners, facilitators, technology intermediaries and researchers,' *Participatory Learning and Action*, 54: 106–113.

Rambaldi, G., Kyem, P. A. K., McCall, M. and Weiner, D. (2006b) 'Participatory spatial information management and communication in developing countries,' *The Electronic Journal of Information Systems in Developing Countries*, 25: 1–9.

Raquez, P. and Lambin, E. F. (2006) 'Conditions for a sustainable land use: Case study evidence', *Journal of Land Use Science*, 1: 109–125.

Reed, M. S. (2008) 'Stakeholder participation for environmental management: A literature review,' *Biological Conservation*, 141: 2417–2431.

Selman, P. (2006) *Planning at the landscape scale*, London: Routledge.

Sieber, R. (2006) 'Public participation geographic information systems: A literature review and framework,' *Annals of the Association of American Geographers*, 96: 491–507.

Stephenson, J. (2008) 'The cultural values model: An integrated approach to values in landscapes,' *Landscape and Urban Planning*, 84: 127–139.

Taylor, J., Murphy, C., Mayes, S., Mwilima, E., Nuulimba, N. and Slater-Jones, S. (2006) 'Land and natural resources mapping by San communities and NGOs: Experiences from Namibia,' *Participatory Learning and Action*, 54: 79–84.

Tuan, Y.-F. (1977) *Space and place: The perspective of experience*, London: Edward Arnold.

Tyrväinen, L., Mäkinen, K. and Schipperjn, J. (2007) 'Tools for mapping social values of urban woodlands and other green areas,' *Landscape and Urban Planning*, 79: 5–19.

URISA Board of Directors (2003) *GIS Code of Ethics*, available at: www.urisa.org/about/ethics#footer%20i (accessed 22 November 2012).

Williams, D. R. and Patterson, M. E. (1996) 'Environmental meaning and ecosystem management: Perspectives from environmental psychology and human geography,' *Society and Natural Resources*, 9: 507–521.

Wood, D. (2010) *Rethinking the power of maps*, New York: The Guilford Press.

Wylie, J. (2007) *Landscape*, London: Routledge.

Zube, E. H. (1987) 'Perceived land use patterns and values,' *Landscape Ecology*, 1: 37–45.

Zurayk, R., el Awar, F., Hamadeh, S., Talhouk, S., Sayegh, C., Chehad, A. B. and al Shab, K. (2001) 'Using indigenous knowledge in land use investigations: A participatory study in a semiarid mountainous region in Lebanon', *Agriculture, Ecosystems and Environment*, 86: 247–262.

15 Seeing both sides: ethical dilemmas of conducting gender-sensitive fieldwork

Experiences from studying perceptions of climate change in Ladakh, India

Virginie Le Masson

> Life is harder for women. Most of the guys they either go trekking or walking around, they don't do much work so women have to do all the work.

Padma said this to me as we squatted around the stove in her kitchen. Of course it was a biased and generalized statement, but perhaps the fact that we were out of earshot of her husband made her feel more confident to say it out loud. Later, my male research assistant interviewed one of Padma's neighbors and asked his opinion as to who usually bore the heaviest workload between the male and female members of his household. He replied:

> Men do more work because they earn, they go in shops and they have to take the main responsibilities. Women they just look after the fields.

I selected Ladakh in the West Himalayan range of Northwest India as my case study location for doctoral field research. I was interested in the perceptions of environmental change among local communities, particularly when compared to other pressing developmental issues such as livelihoods access. My research objectives focused on the gender dimension of climate-change-related discourse, experiences, and practices; thus I wanted to collect data that would reflect both men's and women's perspectives. However, my positionality as a woman had the potential to affect the data collection by biasing my access to participants, their responses, and my analysis. I soon came to the conclusion that for my research to be objective it needed to be gendered.

This chapter explores why I decided to adopt a gender-sensitive approach to fieldwork and the impact of that approach on my research. I discuss the appointment of a gender-balanced research team, the conducting of gender-segregated interviews, and the gender distinctions in interview responses, as well as my choice to feed back the findings in a mixed-gender setting. In each case I reveal some of the ethical issues that I needed to grapple with.

Having gender awareness from the outset

As I was thinking about the most appropriate methodology for my fieldwork, my reading of texts related to cross-gendered and cross-cultural research raised several questions: Are female researchers more legitimized to study women's issues just because they are female? Conversely, is it appropriate for male researchers to study women? What if the cross-cultural dimension is also added: do the lives of a Western female researcher and women in the Global South have enough common ground for them to effectively understand one another's lives?

Gender is a significant aspect of a researcher's identity (see also the chapter by Kovàcs and Bose in this volume), but other personal characteristics such as age, ethnicity, and class also play a role in constructing positionality and influencing the research (see also the chapters by Dam and Lunn and by Godbole in this volume). In researching Asian, middle-aged, and often 'uneducated' women, would it be more significant that I was female, or that I was white and Western, or young and educated? A shared gender does not guarantee easy access to participants and a successful research project: 'Women may have access to other women in the field by virtue of gender, marital status, or childbirth – and men to males – but spatial access does not mean access to the meaning of the worlds of informants' (Warren and Hackney 2000: 6). This statement suggests that gender may not be the overriding factor in shaping my research, yet most of the literature that grounded my work was inspired by studies in gender, development, and the environment and influenced by feminist writings. This literature points out differences in gender roles and status that create inequalities in terms of power and representation but also in relation to people's natural environment. Thus studies often call for gender-sensitive methodologies of data collection whereby women's voices and perspectives are equally acknowledged, valued, and acted upon (e.g., Harcourt 1994; Enarson and Morrow 1998; Masika 2002; ISDR 2008; Dankelman 2010; Fordham 2012).

Having weighed up the literature and the various pros and cons of what my positionality in Ladakh would be, I decided that my approach to fieldwork must be gender-sensitive and that this would entail a gendered-balanced research team, gender-segregated interviews, and a gender-based analysis of responses.

Appointing a gender-balanced research team

My plan was that I would interview female respondents while a male counterpart would conduct interviews with male respondents in each site of study. I had in mind to rely on my partner Krishnan as my male research assistant. He was accompanying me for my fieldwork period in Ladakh and planned to spend the time building up his experience in the development sector (see also the chapter by Lunn and Moscuzza in this volume for more detailed reflections on accompanied fieldwork). With Krishnan acting as my research assistant we could bring our skills and interests together and use this opportunity to design an original gender-balanced way of conducting research. We felt that this arrangement would benefit us both, logistically and emotionally.

Since I can speak neither Hindi nor Ladakhi beyond basic conversational level, I also needed to find an interpreter. And because I had decided on a gendered approach to data collection, that meant I needed two interpreters. I engaged two young university graduates, one male and one female, both of whom spoke English fluently. Although I was all set up and ready to go, my choice to use a gender-balanced team raised several ethical dilemmas about relying on others, delegating research responsibilities, and payment.

During the first period of fieldwork, Krishnan's presence proved extremely valuable on several counts. Nothing went according to plan in the first couple of weeks of the pilot visit. The collaboration that I had set up with a local NGO fell through at the last minute, and this seriously compromised the original objectives of my research. The organization was supposed to help me deal with potential gatekeepers and to facilitate the process of building up contacts with local communities. There was simply nobody to refer to or to rely on anymore. We also had to change our accommodation and ended up paying triple the amount that we had budgeted for.

In the light of this unexpected start I began by arranging informal meetings with other local NGOs in order to introduce myself, present the research, and explore any possibilities to set up a partnership or integrate the research within their work. This process was rather daunting for me, as all of my previous research projects had been conducted in collaboration with a local organization which had facilitated contacts with participants. In contrast, Krishnan was used to conducting research more independently and building himself a network of contacts from scratch. He therefore provided me with a lot of advice, help, and emotional support to find a new approach to my fieldwork.

Hereby arose an ethical dilemma: was it right to rely on someone else for setting up the fieldwork rather than do it on my own? The PhD is a piece of independent research, and with that goes the assumption that the student should deal with any problems or difficulties that arise. However, I found myself dependent on my partner and research assistant to help me cope with a stressful start to the fieldwork. Furthermore, although I was the doctoral student and the research project was mine, I had encouraged our interpreters to stress the fact that we were a team conducting research together, and they usually and spontaneously introduced us as a couple conducting research as part of my PhD studies. This was not because I felt that being a woman in the lead position was culturally inappropriate, nor because I needed to hide behind a male figure to avoid being belittled or rejected, but because I relied on a team from the very beginning to give me strength; this approach also valued the role of my research assistants. So it was more 'us' rather than 'me' leading everyone else. Interviewees as well as our interpreters would often refer questions to any one of us, not necessarily to me. Our presenting ourselves as equals seemed to be perceived positively by participants, and we never encountered anyone who challenged or questioned this way of conducting research. However, it did leave me uncomfortable at times wondering whether it was right that I was not overtly taking the leadership position.

Beyond the initial support that Krishnan provided in developing a new approach and contact network, I also wondered whether it was appropriate that half of the interviews would not be conducted by me, the researcher, but by an assistant. Could the benefits of a gendered approach to data collection really justify the delegation of a major component of my research process?

I had chosen Krishnan not simply because he was my partner and he was at hand but because we shared an academic background, equal qualification, and a similar intellectual approach to the theory which grounded my research. I needed someone whom I could trust with the task of doing half my fieldwork; someone who could understand my research questions exactly, have the confidence to ask questions, and bounce back on answers with the same approach to my own. I knew that Krishnan was suitable in this respect because he had undertaken the same master's degree as me in Disaster Management and Sustainable Development, and thus we subscribed to a similar school of thought.

Although of the same academic mindset, Krishnan brought his own perspective into the interview process, as did our interpreters. When analyzing the data later, I realized how valuable this was for the quality and diversity of the information that we collected. The risk for any researcher using an interpreter is that the interpreter may not translate verbatim everything that is said; for example they may tell you a summary of what someone said rather than the full response, or they may add their own layer of interpretation to the response rather than translate the person's answer. However, I found that my interpreters had a good understanding of the research topic; this allowed for a flexible translation process where interpreters could help interviewees to understand the questions better or help Krishnan and me to grasp interviewees' answers better (see also Leck in this volume for further reflections on working with interpreters). While this certainly improved the quality of the discussions and the data collected, I was still slightly uncomfortable that I was relying on the skill and expertise of other people for the success of my research project.

One could also argue that the delegation of responsibilities and the reliance on others for collecting data may create bias in the interpretation of information. Anthropologists Géraud *et al.* (2007) feel that one should recognize one's subjectivity, acknowledge one's own cultural background and perspective, and accept their constraints on observations and interpretations. Moreover, the involvement of other persons in the research process, not only my research assistants and interpreters but also interviewees and my supervisors, provided subjectivities which have certainly played a role in influencing the interpretation of data. Thus my methodology recognized that potential bias but it assumed the creation of alternative perspectives on real life events based on multiple views, an approach supported by Blaikie (1995).

A further ethical dilemma raised by our gendered research team was the awkward issue of payment. Spending several months in the field comes with associated costs such as accommodation, food, and transport. This circumstance requires either funding to cover these costs or else an assistant willing to pay their own way. From the outset, I agreed to pay my two interpreters a wage and to provide for their accommodation and food when we were out for several days in the field. However, I did not provide Krishnan with a wage or other financial compensation. In effect I benefited from a round-the-clock research assistant for free.

This may sound logical because, as a couple, we shared both earnings and spending; we were also able to share the same accommodation, which saved the cost of additional rent. Furthermore, Krishnan used his savings to pay for his flights. However, this situation still raised the ethical question of whether it was right not to pay my research assistant simply because he was my partner, particularly since I was paying our interpreters a daily wage for the same hours of work. This implied a certain amount of exploitation which lay a little uncomfortably with me. While there was no financial reward for Krishnan this was mitigated by other gains; for example at the time he was looking for a job and was able to use this experience as a research assistant to strengthen his work experience profile.

Overall, my fieldwork was a success because of a partner whose strengths complemented my weaknesses and who was willing to act as my research assistant for 'free,' as well as the contribution of two highly educated and skilled interpreters. I acknowledge a tremendous debt of gratitude to them all but, as explained above, my reliance on others caused me to question how truly 'independent' my research was. In response to this ethical dilemma I would actually like to challenge the widely-held assumption that doctoral research is done alone.

Despite the popular image of the lone researcher, accompanied fieldwork is actually quite common (again see Lunn and Moscuzza in this volume). In fact many of my university lecturers had often suggested that students travelling to areas they are not familiar with or in places that could be dangerous should be accompanied for health and safety reasons. It is also common to see field teams composed of several researchers or PhD candidates conducting their fieldwork as part of a wider research project (e.g., Fagerholm in this volume). To me, this suggests that researchers should not feel that relying on other people – whether supervisors, fellow students, interpreters, friends, or partners – necessarily compromises the seriousness or independence of their research project. I think it is sometimes perfectly justifiable to involve other people in the collection of data or in adaptation to the field setting. As the next section will show, outcomes from my data collection soon helped me to feel confident that involving other people in my fieldwork was the best approach.

Conducting gender-segregated interviews

The purpose of appointing a gender-balanced research team was to carry out gender-segregated interviews. When meeting up with a willing participant, we would often be invited into his/her house to have tea. We wanted to respect the position of the family head (usually a man) so we requested his (or her) participation first. However, because we wanted to show that we valued everybody's opinion, we usually asked another member of the household (usually a female counterpart) whether she (or he) agreed to be interviewed. Krishnan and the male interpreter would then interview a male participant, often in the guest room or in the garden, while I would stay in the kitchen with my female interpreter interviewing one or more female respondents.

Conducting interviews in a gender-segregated environment was significant for my research in three ways. First, it ensured a more inclusive approach. Although Ladakh is often described as a gender-equal society compared to the rest of India, there is still a certain domination exerted by males especially at economic and political levels and in the public sphere; they would normally be put forward to respond to research questions, while women would be expected to look after the visitors by preparing and serving tea. However, our gendered approach meant that women were able to participate in the research while continuing their daily chores (such as cooking or looking after their children) and without it impinging on their time or compromising their responsibilities. Second, the gender-segregated approach resulted in relative openness and honesty in responses because participants were not influenced or contrained by the presence of their female or male kin. Third, it provided me with a means to compare answers between members of the same household or within the same neighborhood and to analyze whether they varied according to gender.

We used a participatory activity to explore gender roles, activities, and perceptions of the resources they owned and controlled within their household. Interviewees were asked whether male or female members of the household were in charge of managing different physical assets (e.g., the television) and daily activities (e.g., collecting water). While some couples had a similar opinion regarding sharing responsibilities, decision-making power and workload, others showed completely different perceptions despite living within the same household. Figure 15.1 shows one example of different perspectives within one household: according to the man (left picture) he was in charge of the fields, fuel

Figure 15.1 Example of a participatory activity conducted with members of the same household to explore their use of and control over their assets; the response shown on the left was by a male participant, and that on the right by a female participant (source: Author, India, 2011).

collection, and managing grain stocks, whereas his wife (right picture) claimed that she was doing these activities.

It is very likely that some of the answers provided by women would have never been given to a male researcher, such as the one cited at the beginning of this chapter. Similarly, it is unlikely that the following statement given by a male interviewee would have been said to me: 'Men should have more power. Otherwise we will be overpowered by the female folks. In Ladakh men should always be on the top, otherwise that will create problems.' My fieldwork experience clearly suggests that the separation of men and women for interviews enabled participants to freely express their own views, reduced the risk of these views being overlooked or hidden due to domination and power structures, and uncovered important differences of opinions between men and women. But inherent in this approach remains the question of whether the shared gender which enabled me to access women and Krishnan to access men truly enabled us to understand their views. The reality was that we were from very different worlds.

> Through gender are transmitted value systems and norms of behavior. This is not to assume that there is one set of norms of behavior characteristic of each gender. What it means to be a male or a female may differ considerably from one social culture to another.
>
> (Oliver 2010: 97)

Once again, it was our two interpreters who made a significant contribution in this respect. After each interview I chatted to my interpreter about my understandings of the interviewee's account; I also discussed them with Krishnan and his interpreter to compare my interpretation with theirs (again see the chapters by Leck and by Lunn and Moscuzza in this volume). As Turner (2010: 212) suggests, the combination of different agencies and identities influenced the social construction of findings; discussions held before and after interviews between the research team members can counterbalance erroneous assumptions and also 'result in far more nuanced understanding of local gender dynamics.'

While these cross-team reflections were very useful in triangulating impressions and findings, they actually presented another ethical dilemma concerning confidentiality. While we felt that women had probably been open and frank in their responses because of the absence of their male kin, I sometimes discussed their responses with Krishnan and his interpreter who had been interviewing those very male kin, in order to compare participants' views and better grasp the family contexts. However, I wondered whether this violated the confidentiality of participants' responses. One of the benefits of our gendered approach, as noted above, was that it had the potential to give voice to members of a household or community who are normally marginalized, secluded, or oppressed. To conduct research with such people thus often carries additional ethical requirements concerning confidentiality in order to protect them from harm. I would have not shared any information with my male research assistants if there had been a risk to the wellbeing of female participants, but I felt that none of the information I was obtaining was taboo or sensitive enough to conceal from my research assistants. Besides, the very process of comparing findings served to highlight differences of opinion and to raise awareness, even among our research team, that a gendered approach matters when exploring environmental changes. This led me to wonder whether such gender awareness should not also be created among participants themselves. This reflection informed the final part of my fieldwork, as I shall describe in the next section.

A non-gendered approach to feeding back

My research did uncover a difference between men and women in their perceptions and experiences of environmental changes, their gendered vulnerabilities and capacities, and their opportunities to be involved in decision-making related to development and climate change that affected their everyday lives.

It is good ethical practice to feed back the overall research findings to participants, since they are the owners and creators of the knowledge. As with the data collection, I could have chosen a gendered-approach to feeding back, telling each group about the differences between male and female responses. But I felt that this separation would actually serve to reinforce the gender divide within those communities and would not challenge the structures of domination mainly exerted by men. In my view, both male and female participants needed to sit together in order to recognize that gender does matter when exploring environmental issues.

But the ethical dilemma lay in whether it was my place as an outsider to challenge the status quo. I had observed socio-cultural structures that in my opinion were not 'right' and which maintained certain groups in more oppressed positions in that society. We are present in the field as independent researchers and with that goes the assumption that we are subjective, unbiased, and detached. However, the reality is that we are human beings with morals, feelings, and opinions. Many of us are involved in development studies because we have some kind of moral conscience urging us to strive to make the world a better place for those who are poor, underprivileged, and oppressed. So why do research to better understand the world without trying to change it? This debate also revolves around the idea of justice, what it means for people, and who should take the responsibility for fostering justice in the development of societies. Thus I felt strongly the need to use my research findings to challenge the inequalities and vulnerabilities that I had observed.

I returned to the field for a third time in order to organize focus-group discussions with both male and female participants. For me the overt purpose was to triangulate the data; for them it was an opportunity to reflect on their participation in the research, to be actively involved in the discussion of the findings, and to express their opinions regarding both the methodology and the topic. I ensured that there were an equal number of men and women in each group and the resulting debates enabled the participants (and me) to recognize differences of views between men and women. But beyond the importance of highlighting differences it provided a platform for both men and women to voice their particular views and defend them from an equal stance. Eventually, participants extracted points of convergence which they used to reach a consensus on specific topics.

Therefore, the research process I adopted did not just separate men and women to guarantee their equal participation but eventually and voluntarily put them together to reach decisions that were mutually constructed despite differences. I am aware, though, that what worked in the context of Ladakh might not be possible in other contexts. Other researchers would need to consider carefully the best way to foster inclusiveness, equal representation, and empowerment.

Some final reflections on a gendered approach to fieldwork

I decided that the gender dimension of my research questions was best addressed by a gender-balanced team conducting gender-segregated interviews that would collect gender-specific data. This chapter has reflected upon some of the ethical dilemmas of this approach. My concerns about relying on others and delegating research responsibilities were mitigated when it became clear that the gendered approach ensured the inclusivity of both male and female respondents and created a space where participants could talk about gender-related issues in a more confident way, particularly in relation to roles and power.

Despite the benefits of conducting gender-segregated interviews, it was not a guarantee that the shared gender automatically meant privileged access into the participants' worlds. In this respect our interpreters were instrumental in crosscutting the bias of our positionalities with different ethnic backgrounds and other perspectives. While the gender-specific responses were very significant to my research, I did not want my approach to reinforce the gender divide among local communities. Thus my organization of mixed focus-group discussions created a platform for women and men to voice their particular views and defend them from an equal stance before reaching decisions that were mutually constructed.

While the gendered approach was a great success for my research, it left me with some unresolved ethical dilemmas with regards to my relationship with my research assistant. Quite simply I would not have been able to undertake the same research without the help, skills, and support of Krishnan; conversely I would not have been able to afford the same services had he not been my partner. This questions the possibility for non-relational couples to adopt a similar methodology unless there is a budget to pay the assistant or unless he or she is willing to work for free. I hope that my confessions and reflections have debunked the assumptions based on the concept of the lone researcher and have emphasized that successful 'independent' research can by carried out with the help of research assistants and interpreters.

Acknowledgments

I would like to express my deepest gratitude to my partner and research assistant, Krishnan Nair, for his participation in my fieldwork and his contributions to drafts of this chapter. My thanks also go to our interpreters, Stanzin Angmo and Rigzin Angchuck, as well as to Tsewang Dolma, for their help in the field and valuable inputs into my research.

Recommended reading

Enarson, E. (2003) *Working with women at risk: Practical guidelines for assessing local disaster risk*, Miami, FL: International Hurricane Center, Florida International University, available at: www.ihrc.fiu.edu/lssr/workingwithwomen.pdf.

This text provided me with concrete advice and tips for setting up a gender-balanced research team and a gender-sensitive research process.

Leduc, B. (2009) *Guidelines for gender sensitive research*, Kathmandu: International Centre for Integrated Mountain Development (ICIMOD), available at: www.icimod.org/resource/1288.

This short guide note offers a useful synthesis on gender-sensitive research, emphasizing the reasons gender matters and the ways gender can be integrated into participatory research.

Pincha, C. (2008) *Gender sensitive disaster management: A toolkit for practitioners*, Mumbai: Earthworm Books, available at: http://gdnonline.org/resources/Pincha_GenderSensitiveDisasterManagement_Toolkit.pdf.

This toolkit offered numerous constructive ideas and examples of participatory activities that can be used to explore gender differences when studying vulnerability and the capacity to face hazards. It was particularly relevant for the Indian context.

Warren, C. and Hackney, J. (2000) *Gender issues in ethnography*, 2nd edn, Sage University Papers Series on Qualitative Research Methods, Vol. 9, Thousand Oaks, CA: Sage.

This short book was particularly interesting and I found it accessible for non-anthropologists/non-ethnographers such as myself. It offers a useful synthesis of studies that address gender issues when conducting ethnographic fieldwork. Several sections echoed my experience of doing research with my partner, and there were accounts of other types of cross-gender teams.

References

Blaikie, P. (1995) 'Environments or changing views? A political ecology for developing countries,' *Geography*, 80 (3): 203–214.
Dankelman, I. (2010) *Gender and climate change: An introduction*, London: Earthscan.
Enarson, E. and Morrow, B. H. (1998) *The gendered terrain of disaster: Through women's eyes*, Miami, FL: Laboratory for Social and Behavioral Research, Florida International University.

Fordham, M. (2012) 'Gender, sexuality and disaster,' in B. Wisner, J. C. Gaillard and I. Kelman (eds), *Handbook of hazards and disaster risk reduction*, London: Routledge.

Géraud, M., Leservoisier, O. and Pottier, R. (2007) *Les notions clés de l'ethnologie: Analyses et textes*, Paris: Armand Colin.

Harcourt, W. (1994) *Feminist perspectives on sustainable development*, London: Zed Books.

ISDR (United Nations International Strategy for Disaster Reduction) (2008) *Gender perspectives: Integrating disaster risk reduction into climate change adaptation – Good practices and lessons learned*, Geneva: United Nations.

Masika, R. (2002) *Gender, development, and climate change*, Oxford: Oxfam GB.

Oliver, P. (2010) *The Student's Guide to Research Ethics*, 2nd edn, Maidenhead: Open University Press.

Turner, S. (2010) 'Research note: The silenced assistant – Reflections of invisible interpreters and research assistants,' *Asia Pacific Viewpoint*, 51 (2): 206–219.

Warren, C. and Hackney, J. (2000) *Gender issues in ethnography*, 2nd edn, Sage University Papers Series on Qualitative Research Methods, Vol. 9, Thousand Oaks, CA: Sage.

Part IV

Ethical dilemmas of engagement

16 'You can be jailed here by even me talking to you': dilemmas and difficulties relating to informed consent, confidentiality and anonymity

Experiences from field research into violence in Israel and Palestine

Chloe Skinner

[My husband] is still controlling now … If he hear me [talking to you] he is very angry. If he hear me, he is very angry. There is many, many times he beat me, he has broken my arm.

(Palestinian woman living in occupied West Bank)

This quote from a participant in my doctoral research, and those included elsewhere in this chapter, illustrate the very real context of fear and violence in which I conducted fieldwork in Israel and Palestine. The sensitivity of my research topic necessitated adherence to the principles of informed consent, confidentiality and anonymity. While these principles are an important consideration for all researchers, fieldwork in certain contexts makes applying them particularly necessary but also tricky; for example working in dangerous countries, researching sensitive topics, or working with vulnerable groups such as children (see also the chapters by Brooks, Day, Taylor, and Tomei in this volume).

My doctoral research aimed to explore masculinities in the context of violence in both Israel and occupied Palestine and the ways in which masculinity formation and negotiation in this context is perceived to relate to violence against women, both Israeli and Palestinian. Protracted conflict, violence, and military occupation have shaped the everyday lives of both Israelis and Palestinians, in differing ways, since the formation of Israel as a state in May 1948 (known as 'al-Nakba,' meaning the catastrophe, for Palestinians and 'Independence Day' for Israeli Jews). Yiftachel and Ghanem (2004) contend that Israel now functions as an 'ethnocracy' through which ethnicity is prioritized over inclusive citizenship, facilitating the ethnicization, expansion, and control of one dominant ethno-group over contested territories and power structures, thus creating political and military structures which ensure a uni-ethnic seizure and control of a bi-ethnic state (see Yiftachel 2006).

With villages and towns pulverized, olive groves uprooted, houses demolished, and reservoirs destroyed, Palestinians living in Israel, Gaza, and the West

Bank have been enclosed under the 'matrix of control' of occupation through the construction of an eight-meter-high wall, the imposition of hundreds of check-points, and an 'ethnocratic' regime under which they are severely subjugated (Lagerquist 2004; Yiftachel and Ghanem 2004; Pappé 2007; Roy 2007; Weizman 2007; Halper 2012). This functions to 'send a message of force and authority, to inspire fear, and to symbolize the downtrodden nature and inferior-ity of those under occupation,' says the former mayor of Jerusalem (Benvenisti, quoted in *Haaretz* 2002: 1; see also Gregory 2004; Peteet 2002).

Meanwhile, my research has led me to believe that Israeli Jews are also subject to forms of violence by the Israeli state through near universal conscription to the military and the discursive creation of a context of fear, xenophobia, and paranoia with regard to all neighboring states, exacerbated by a history of violent anti-Semitism in Europe and elsewhere, and acts of violent resistance by some Palestin-ians. This contributes to the wholesale militarization of Israeli society, shaping childhood, family life, time, public space, politics, the economy, and, as demon-strated through my research, gender and sexualities (Adelman 2003).

In such a context, as for all research, the maintenance of the principles of informed consent, confidentiality and anonymity are fundamentally important; yet in this context they are not always easy to achieve clearly, as I will illustrate in this chapter. I draw from my fieldwork experiences involving in-depth inter-views with 29 people including both male and female former and current Israeli soldiers, and male and female Palestinians living under military occupation in East Jerusalem and the West Bank.

Corti *et al.* (2000: 3) write that qualitative research 'should, as far as possible, be based on participants' freely volunteered *informed consent*. This implies a responsibility to explain fully and meaningfully what the research is about and how it will be disseminated.' This exercise, in which participants are norma-tively given autonomy in the research process, principally exists to diminish the potential for the exploitation of human participants (Rosenthal 2006). Yet Corti *et al.* (2000) explain that the extent to which participants can ever be fully 'informed' is disputed. This is, perhaps, particularly the case for research with 'subaltern' participants whose subjugation on the basis of class, geography, race, and gender or other factors may render them unable to be 'spoken for' through the research process (Spivak 1988) as well as unable to access fully the 'informa-tion' necessary to give 'informed' consent.

In relation to confidentiality and anonymity, the statement of practice of the Developing Areas Research Group (DARG 2009: 1) says that 'It is imperative that the identities of respondents are protected in discussions, presentations and data sets, and thus appropriate mechanisms should be deployed to guarantee con-fidentiality and anonymity.' According to DARG this is a key 'right' of partici-pants, and 'because of the potential risks associated with revealing respondents' true identities, no other consideration should be allowed to supersede this funda-mental principle.' Thus, particularly in the context of conflict, an authoritarian regime, violence in the home and elsewhere, and in other dangerous circum-stances, the identity of participants must be concealed in order to prevent harm coming to those involved in research projects. Yet, as outlined below, the DARG statement of practice is not always easy to achieve.

I would like to share four 'moments' during my doctoral research in which these essential ethical considerations were difficult to uphold. The first relates to informed consent and the 'signing of forms' to signify this; the second involves the tension between the expression of researcher emotions and confidentiality; the third relates to the fear of passing through Israeli airport security with my research data and ways in which I attempted to address this; and the fourth involves confidentiality and informed consent with regard to research dissemination.

Informed consent: to sign or to speak?

Prior to fieldwork I felt pressure to assert that I would gain written informed consent from participants. It seemed to be the 'most ethical' approach and, in truth, would enable me to get through the bureaucracy of the pre-fieldwork ethical review process with my challenging research proposal. I prepared a carefully-worded form for participants to read and sign which would verify that they understood the nature of the research and their rights relating to participation, including their rights to withdraw, to anonymity, and to confidentiality. However, upon starting fieldwork I soon discovered that while this proved helpful and informative for some participants, for others the gathering of 'written informed consent' went against one of my university's general principles of ethical research which states that 'cultural sensitivity' is a fundamental obligation of the researcher (University of Sheffield 2013: 2).

Israeli participants, all but one of whom had served in varying capacities in the Israeli military or national service (ranging from officers in combat units to presenters on the official radio station of the Israeli army), appeared comfortable and even pleased to read and sign an informed consent form. For example, one participant wished to see the form before meeting me and expressed that he felt assured to see that I was legally committed to his confidentiality and that I was 'doing it properly.' Yet these participants, privileged with equal citizenship within Israel's ethnocratic structure not only as Jews but also as Jews who have served their national or military service, have not experienced explicit systemic marginalization under the Israeli regime, which is reserved for Palestinians and 'other' ethnic groups in Israel and the occupied Palestinian territories, and, in a differing way, Israeli Jews who have refused to serve in the military. Rather, Israeli state systems and structures of 'legality' and 'bureaucracy' have functioned to legitimize their actions and 'privileged' ethnic status in society. Thus the presentation of a form or a contract in my research appeared either comforting or of little concern to these participants.

On the other hand, Palestinian participants to whom I presented an informed consent form appeared cautious at best, if not suspicious. This is unsurprising given their subjugation under the laws and bureaucracy of Israel (Yiftachel 2006) through which one step out of a very narrow box could lead to their imprisonment, the demolition of their house, or other such punitive measures.

One participant informed me of a moment in his neighborhood in which elderly Palestinians signed consent forms (in Hebrew) for the demolition of their own homes; he explained that for this reason children now learn Hebrew in his neighborhood. The level of psychological distress that I potentially could have caused to Palestinian participants through my presentation of informed consent forms is clear in such a context, despite having translated the forms into Arabic.

During my fieldwork, I felt confused about whether to continue asking Palestinian participants to at least read – if not sign – the forms. I settled on the decision to mention that I had a form, the contents of which I would explain to them verbally if they would prefer. In hindsight this confusion was not necessary and the answer is now clear to me: in such a setting it is unethical, culturally and politically insensitive, and potentially distressing to ask such individuals to read and sign a legal document purely in order to present a full set of completed forms to my institution should they request it; the gaining of verbal informed consent is quite sufficient. Although I came home with an incomplete set of informed consent forms, trustingly signed by all Israeli participants and cautiously signed by some Palestinian participants, should my examiners or university ethics board request to see these, I feel confident in explaining the reasons for my approach.

Balancing emotional catharsis and confidentiality

> I went through a serious sexual assault as a kid, and I have never told my boyfriend. I think that is where it is really important to talk, but it is a hard thing to do, it is just like ... you don't want to open up, and you don't want to tell anyone ... I never told anyone. I never told any men I was in a relationship with. I have told my therapist, and my best friend.... It was my uncle, so that's why. My uncle assaulted me as a kid. For a few years. And I have never told anyone in my family.... They still don't know, no one. But like I feel that it is something that I chose. And I am very ... I am happy that that was my decision, because I feel like it would break up my family. And my family is ... I am very close to my family, and it is very important to me ... so I feel like it is a price I am willing to pay.
>
> (Israeli woman, former non-combatant soldier)

Research can be emotionally draining and distressing for the researcher (see also the chapters by Day and Smith in this volume). This can further complicate ethical considerations as it can be difficult to separate ethics from emotions (Meth and Malaza 2003). As illustrated in the quote above, I was working with participants who had experienced severe traumas such as rape, domestic violence, murder, imprisonment, and physical beatings, as well as fear, paranoia, poverty, spatial restriction, and insecure housing. As well as victims I was also working with perpetrators of some of these acts of violence. Add to this the emotional stress of carrying out this research in a dangerous, or at least threatening, context (again see chapters by Brooks, Taylor, and Tomei in this volume).

In the case above, I was one of the very few people to be told about some of these horrific experiences. At best I felt helpless; at worst I felt voyeuristic. Yet one of the aims of my research was to contribute to a wider body of feminist academic and activist work which breaks the silence surrounding violence against women. Thus I, with the help of others, reconciled myself that being told such experiences was necessary for my work, hopefully helpful for those doing the retelling, and an important step in the wider project of exposing and challenging the presence and effects of violence against women.

During fieldwork and throughout the subsequent process of transcription and analysis, I felt the need to emotionally 'off-load' to my close friends, family, and supervisors. At times discussion of my research has led to an abrupt end to the conversation. I am certainly not alone in this: Etherington (2000: 2) writes that 'when people asked "what is the subject of your research?" and I told them, the conversation stopped.' I have found the solitary – and at times isolating – nature of independent research a challenge, more so in the context of research into violence. Thus when given an opportunity to speak of the content of my research and my emotions about it, I have felt a catharsis, a release of pressure and distress.

Yet this cathartic process of emotional 'off-loading' requires divulging and discussing participants' experiences and traumas in a way which may not be as carefully thought out as in a written academic document. There were times when I questioned whether I had revealed too much. Had I discussed other people's experiences too openly? Had I represented my participants fairly? Had I depressed my listeners? Was I prioritizing my emotions and emotional needs over research ethics? I felt as if I had perhaps crossed the line of confidentiality in a bid to emotionally 'off-load' for my own sanity. Of course I maintained the anonymity of participants, yet I was concerned that I had too openly, and without careful wording and consideration of my representation of participants, discussed contents of interviews without their explicit consent. While this process of emotional expression was necessary for me, there are tensions between emotional catharsis and participant confidentiality. This very necessary process of informal 'data dissemination' illustrates the blurring of boundaries between how data is 'used' in a way which is not fully recognized in formal ethics review processes.

Ethical research in a surveillance state?

> When the whole society is more violent ... you can see people here are more edgy, so it has to come out. So it has an effect, it is now with your research.... Are they hard [on you] at the checkpoints?... Wait till you go to the airport. Be ready for it I think, I don't want to make you nervous. But it is not so bad, if you take it somehow without context. You will be checked, it is not against you, it is written in a book. People travelling alone and also younger people.
>
> (Male reservist officer in an Israeli combat unit)

You can be jailed here by even me talking to you, so it's tough.... Look, make sure nobody hears [this] at the airport – they can take this from you ... they are really strict about this stuff you know.

(Palestinian man living in occupied West Bank)

The 'architecture of occupation' sustained by the Israeli state in both Israel and the occupied Palestinian territories is dependent upon surveillance, bolstering Israel's capacity to govern (in differing ways) the Palestinian population, Israeli citizens, and visitors to Israel and Palestine (Weizman 2007; Zureik *et al.* 2010). Imposed through checkpoints, military watchtowers, ID cards, color-coded paperwork for Palestinians, spatial restrictions, unmanned aerial vehicles, and innumerous other techniques, this enables the state to determine which bodies are where, to monitor and control those designated as threats, and to control the mobility, and indeed minds, of those subjected to this biopolitical system of governance, as soldiers, as occupied bodies, as dissidents, and so on (Zureik *et al.* 2010).

This control is not only physical but also psychological. The sense of fear and threat on passing through the criminalizing checkpoints is palpable; worry, anger, and apprehension is etched onto Palestinian faces as they wait in line in areas resembling cages and cattle sheds while bored young Israeli soldiers – M16s casually swinging – ignore, order, search, or detain, and, ultimately, control those who 'dare' to pass.

At one particular checkpoint – Qalandia – from Jerusalem to Ramallah, passports of international people are often scanned. When I had to pass through here I had a sense of being watched and was worried about the future implications of my passport being scanned when I came to leave Israel via Tel Aviv airport. On one such occasion, my passport was taken away for ten minutes to be scanned and checked while the young Israeli soldier in front of me barked orders and questions between performances of 'official phone calls.' There were no apparent consequences to this event; after she threw my passport back telling me to 'take it and get lost,' no questions were asked about my supposedly 'scanned' passport at the airport, and no follow-up commenced after her phone calls; yet the psychological effect was significant. Who did she phone? What would I say at the airport? What if they asked for my data?

In the context of military occupation, the occupier – the Israeli state – presents itself as the omnipresent, all-seeing, all-knowing entity with eyes and ears everywhere. For myself, as for other international researchers and activists, and of course predominantly for Palestinians, this is a constant source of psychological distress, functioning to influence the actions of the people moving through checkpoints and airports and passing under military watchtowers. As illustrated in the quotes above, two men – one Israeli and one Palestinian – warned me to take care at the airport and to ensure that nobody heard my interview recordings.

I felt panicky as my departure date drew closer, knowing that the contents of some interviews could potentially have negative consequences for interviewees should they fall into the wrong hands. I had stored my transcripts and recordings online and had emailed many to myself, but I did not feel fully secure with this so had further copies on a memory stick. After seeking advice from locals and other international visitors I decided to mail the memory stick home rather than carry it through the airport where I would be checked rigorously.

Three weeks later my package arrived at home. It included detailed maps of the occupied territories and illegal settlements, books by dissenting Palestinian and Jewish authors and organizations, drawings by some participants, and various other items which would have needed 'explaining' should it be checked by airport security. Missing from the package was my memory stick. I imagined the worst. I couldn't rest easy for weeks and deliberated about getting in contact with participants to inform them of what had happened. I decided that this would cause substantial psychological harm and so did not get in touch, all the time waiting for the worst news.

As I write this chapter eight months later, nothing has happened; none of the participants with whom I maintain contact have mentioned any repercussions, and I have received no news from others who have my contact details. Yet still I worry about the whereabouts of the memory stick. I feel comforted by the fact that all of my other mailed items were received, and rational voices have questioned why, if the package had been seized by the state, would everything else have been carefully packaged and sent on with only the memory stick confiscated. Ninety-nine percent of the time I feel sure that the memory stick – as the smallest item in the package yet the one which was relatively heavy in comparison to the papers – must have fallen out of a hole and been destroyed, stolen or broken. But a small part of me continues to fret and question what else could have happened, indicating the psychological power of the disciplinary techniques of the Israeli state.

Informed consent, research dissemination, and 'spectacle'

Following six months of analyzing my fieldwork data I gave a presentation at a conference, the first since completing my fieldwork. This first occasion of public dissemination made me question the 'informed' nature of 'informed consent.' I was verbally quoting participants, specifically male Israeli soldiers. I wondered whether they had been really aware that they might be directly quoted by me later in the course of dissecting, analyzing, questioning, and presenting components of my research.

Despite ensuring the anonymity of these participants through the use of pseudonyms and attempting to ensure fair representation of each of them, I still felt discomfort in using quotes from our interviews. Was I making a spectacle of those who had participated in the research? I have and will continue to quote participants as this is necessary for the dissemination of qualitative research. However, it felt uncomfortable, particularly in the context of a conference rather than in written format. Perhaps it was because I was saying their words to an audience whose reactions – often visible when discussing violence – I could see; in written dissemination the 'audience' – the reader – is distant and the researcher is removed from their reactions. The experience of witnessing visible reactions from an audience when presenting qualitative research has led me to question how informed 'informed consent' really is and has emphasized the utmost importance of fair and complex representation of participants.

(An un-concluding) conclusion

My experiences and reflections on trying to adhere to informed consent, confidentiality and anonymity during fieldwork have shown that many of my ethical dilemmas remain unresolved. I am painfully aware of the 'gray areas' of ethical research practice. Somewhat prone to guilt, I have found it difficult to rest easy with these dilemmas, which have made me question whether research can ever be truly ethical in an unethical world. It seems to me that the distress related to these 'gray areas' could be termed an occupational hazard of research, particularly for those working in certain contexts such as with vulnerable participants, with victims or perpetrators of violence, in volatile places, in the Global South, or when researching sensitive subjects.

I have found that sharing experiences and difficulties through honest and open reflection with other research students, my supervisors, family, and friends – and writing this chapter – has really helped to negotiate this 'occupational hazard.' I hope that this chapter will encourage other researchers to be open with those around them about their dilemmas and difficulties and to engage in honest conversation around ethical research practices (and divergences from them). I also hope that my experiences will encourage others to be as aware of their research context as possible when deciding whether to gain written or verbal informed consent and when considering the transportation of data during and following fieldwork. Finally, in the process of research dissemination, I suggest that while 'interesting' research is important, it is essential to be reflective about why certain quotes from participants are included or not; if it is merely to 'grab the attention of an audience' it is perhaps better left out!

I continue to search for a place where my conscience is comfortable, if not necessarily clear. If such a place cannot be found, surely ethically challenging research such as mine, which seeks to explore in order to contest myriad forms of violence, would cease to exist.

Recommended reading

Corti, L., Day, A. and Blackhouse, G. (2000) 'Confidentiality and informed consent: Issues for consideration in the preservation of and provision of access to qualitative data archives,' *Forum Qualitative Sozialforschung/ Forum: Qualitative Social Research*, 1 (3), art. 7.

This article outlines some key debates relating to informed consent and confidentiality in relation to qualitative research. For me, although it is not directly related to in-depth interviews, this served as a helpful entry point to considering these ethical principles.

Meth, P. and Malaza, K. (2003) 'Violent research: The ethics and emotions of doing research with women in South Africa,' *Ethics, Place and Environment,* 6: 143–159.

This open and honest reflection on researching violence in a violent context questions the blurry boundaries between ethics and emotions. This article will be helpful to those conducting challenging and emotionally distressing work in a difficult context.

References

Adelman, M. (2003) 'The military, militarism, and the militarization of domestic violence,' *Violence against Women,* 9: 1118–1152.
Corti, L., Day, A. and Blackhouse, G. (2000) 'Confidentiality and informed consent: Issues for consideration in the preservation of and provision of access to qualitative data archives,' *Forum Qualitative Sozialforschung/Forum: Qualitative Social Research,* 1 (3), art. 7.
DARG (Developing Areas Research Group) (2009) *DARG ethical guidelines,* available at: www.devgeorg.org.uk/?page_id=799 (accessed 10 May 2013).
Etherington, K. (2000) 'Supervising counsellors who work with survivors of childhood sexual abuse,' *Counselling Psychology Quarterly,* 13 (4): 377–389.
Gregory, D. (2004) *The colonial present: Afghanistan, Palestine, and Iraq,* Malden, MA: Blackwell.
Haaretz (2002) *The checkpoints of arrogance,* available at: www.haaretz.com/print-edition/opinion/the-checkpoints-of-arrogance-1.51106 (accessed 1 April 2012).
Halper, J. (2012) *The key to peace: Dismantling the matrix of control,* available at: http://uk.icahd.org/articles.asp?menu=6&submenu=3 (accessed 23 February 2012).
Lagerquist, P. (2004) 'Fencing the last sky: Excavating Palestine after Israel's "Separation wall",' *Journal of Palestine Studies,* 33: 5–35.
Meth, P. and Malaza, K. (2003) 'Violent research: The ethics and emotions of doing research with women in South Africa,' *Ethics, Place and Environment,* 6: 143–159.
Pappé, I. (2007) *The Israel/Palestine question: A reader,* London: Routledge.
Peteet, J. (2002) 'Male gender and rituals of resistance in the Palestinian Intifada,' in R. Adams and D. Savran (eds), *The masculinity studies reader,* Oxford: Blackwell.
Rosenthal, J. (2006) 'Politics, culture and governance in the development of prior informed consent in indigenous communities,' *Current Anthropology,* 47: 119–142.
Roy, S. (2007) 'Humanism, scholarship, and politics: Writing on the Palestinian–Israeli conflict,' *Journal of Palestine Studies,* 36: 54–65.
Spivak, G. (1988) 'Can the subaltern speak?,' in C. Nelson and L. Grossberg (eds), *Marxism and interpretation of culture,* Chicago: University of Illinois Press.
University of Sheffield (2013) *Research ethics: General principles and statements,* available at: www.shef.ac.uk/polopoly_fs/1.112655!/file/General-Principles-and-Statements.pdf (accessed 10 May 2013).
Weizman, E. (2007) *Hollow land: Israel's architecture of occupation,* London; Verso.
Yiftachel, O. (2006) *Ethnocracy: Land and identity politics in Israel/Palestine,* Philadelphia; University of Pennsylvania Press.
Yiftachel, O. and Ghanem, A. (2004) 'Understanding "ethnocratic" regimes: The politics of seizing contested territories,' *Political Geography,* 23: 647–676.
Zureik, E., Lyon, D. and Abu-Laban, Y. (2010) *Surveillance and control in Israel/Palestine,* Abingdon: Routledge.

17 Giving the vulnerable a voice: ethical considerations when conducting research with children and young people

Experiences from fieldwork in Zambia

Caroline Day

At just 14, Betty has already experienced so much. As head of the household she is responsible for her whole family. Not only does she have to look after her sick mother, whose frequent fits and obvious confusion would be a challenge for any adult, but she also has to care for her two younger sisters too and try to earn an income to ensure they are all fed and clothed while trying to also afford the medicines her mother needs. On top of this she is also pregnant! It seems crazy that with her being under the age of 16 I still had to ask for her mother's consent to speak to her when she is the one that makes decisions every day, not only for herself but for her whole family, but this I duly did, as well as ensuring she gave her own consent too.

    She sat across from me, tears welling up in her eyes as she described her life. She was hungry, stressed and felt totally betrayed by the boy who she thought had loved her. On top of this, her mother's anger at her pregnancy was palpable. I felt guilty trying to probe her about how she was coping, but she insisted she wanted to talk. I did not want to cause her any further distress, but at the same time I think I was probably the first person to listen to her story from her perspective, listen to her fear that her mother would die and she would be left bringing up her siblings and a baby single-handedly.

This incident captures some of the issues that I faced when conducting research with children and young people in Zambia looking at how their transitions to adulthood are affected when they have caring responsibilities for sick or disabled parents or relatives. Doing research about and with children and young people is fraught with added layers of complexity and sensitivity compared to research with adults, due to their vulnerability.

Since the early 1990s there has been a burgeoning of research in the social sciences on children and young people (Evans 2008), with particularly valuable contributions coming from the field of geography (Holloway and Valentine 2000). The recognition that children and young people are social actors (Duckett *et al.* 2008) who are 'competent witnesses to speak for themselves about their experiences of, and perspectives on, the social worlds in which they live' (Barker and Weller 2003: 208) has led to greater recognition that children and young people not only have the 'right to express their views freely' but 'have valuable knowledge and understanding' (Robson *et al.* 2009: 470) about the issues affecting their lives. Such research is based on being 'fair and respectful' to young people and recognizes that they are research 'subjects' rather than 'objects' (Barker and Weller 2003; Ansell 2005; Greene and Hill 2005; Clacherty and Donald 2007) who rather than having research done 'on them' can actively participate and contribute to research studies 'about them.'

However, with this has grown the recognition that while children and young people can offer a valuable contribution to research, they also occupy a vulnerable and powerless position in society in which they can be manipulated (Morrow and Richards 1996; James *et al.* 1998; Masson 2000; Evans and Becker 2009). While research can give young people a voice, it can also be an 'intrusive process' in which young people are asked to discuss sensitive issues (Lindsay 2000).

Ethical considerations are relevant to research regardless of the age of the participant, but it is widely recognized that they need to be more stringent when conducting research with children and young people compared to adults (Scott 2000) and must offer greater protection in research relationships. Therefore, while participation of children and young people in research has been actively encouraged, it has been on the understanding that it should not be at the expense of their protection from additional harm or distress (Evans and Becker 2009). Ethics has therefore become a key part of researching youth, 'predicated on the expectation that the participants [will] suffer no harm as a result of the research or its outcomes' (Young and Barrett 2001: 130).

This chapter explores my experiences in Zambia, focusing on four elements of my fieldwork in which ethical considerations were a key feature of the planning and execution: identifying the research participants; gaining informed consent and ensuring participant safety; selecting the research methods; and recognizing young people's participation. I illustrate how ethical considerations influenced my behavior and research practice in the field, as well as some of the challenges that occurred and how I overcame them.

An ethical approach from the outset

In deciding to work with young people in my own research, I made ethical considerations a part of my planning from day one. Such planning was also a necessity as I had to have ethical approval granted to me in advance of starting the fieldwork, not only from the ethics committee at my university but also from the University of Zambia (UNZA), with which I would be affiliated while I conducted my fieldwork.

For my own university ethics committee I had to submit a full application, including project submission form, a research proposal, and copies of all my interview and focus group schedules, information leaflets, and consent forms. I had to complete this some months in advance to ensure that ethical approval had been given before I commenced the fieldwork. My submission was accepted, but

I was asked for a small number of clarifications in the detail as well as for greater assurance of my safety as a lone female researcher in a 'developing' country. I was also told that I was prohibited from enquiring what the illness or disability was of the parent or relative that each young person was caring for, as I did not have a medical background. This was not a huge hurdle as my research focus was the young people's transitions, not specifically their caring role on a day-to-day basis, but in other circumstances this could have proved tricky.

The University of Zambia required a similar submission of documents to their own ethics committee to enable them to decide whether to allow me to undertake the research and to therefore agree to become my affiliate institute and aid my application for a residence permit. This I undertook while in Zambia during a pilot month prior to the main fieldwork taking place. As I was an international student, the processing of my application required a substantial cash payment, while limited resources at the university as well as a restricted and unreliable internet connection meant that the research proposal, interview schedules, and all other documents had to be submitted in hard copy 15 times over for distribution to each member of the committee – something that it was definitely useful to be 'in-country' at the time to achieve. I was then able to leave the application with the committee knowing that I was not due back in the country for another five months, within which time it was successfully approved.

Having to consider ethical issues so early on meant that I had to consider and decide various aspects of my research process long in advance of the fieldwork actually taking place. This did not mean that ethical challenges would not arise while in the field, but it did ensure that key parameters were already in place to guide my behavior and ensure the safety of both the participants and myself as soon as my fieldwork began.

Ethical challenges of participant recruitment

The young people involved in this study were actually older 'children' and young people aged from 14 to 30, rather than young children, on which there has been much dialogue about ethics (Alderson 1995; Morrow and Richards 1996; James *et al.* 1998; Scott 2000; Barker and Weller 2003; Alderson and Morrow 2004; Van Blerk 2006). My decision to work with this age group was predicated on the fact that my research was to focus on 'transitions to adulthood' and therefore the life experiences that young people closer to becoming 'adult' were currently undergoing. It was also in recognition that much of the research on young caregivers so far had tended to focus on children and the day-to-day experiences of caring, rather than on older youth and how caring roles both influence and impact their futures and life transitions (although there are exceptions to this, e.g., Evans and Becker 2009; Evans 2011). It has been argued that this reflects a general focus in geography on the 7–14 age group while the 'discipline has been slower to consider young people on the cusp of childhood and adulthood: those aged 16–25' (Valentine 2003: 39).

While only a proportion of participants in my sample would fall under the age of 18 and therefore be classed as 'children,' I still had to take into account the social, cultural, and economic factors that would emerge during the research, particularly my position as an educated white female and the power relations that this can entail (Evans and Becker 2009). I did not want the participants to feel in any way obliged to participate, nor to feel that they were required to give particular answers or responses to me as the researcher based on what I wanted to hear. I therefore applied the same level of ethical consideration to all the young people, irrespective of age, education, or social status, so that they all had the same information prior to consenting to take part.

The young people were identified with the support of two key organizations based in the Lusaka area that worked with families with diverse health and care needs, as well as families where youths were not caring for someone sick or disabled. Both organizations had a number of staff or volunteers who worked within the communities in which I wanted to be based and so already had relationships with young people and families who fitted my research criteria. This enabled me to spend time in the communities prior to recruiting participants.

By being a physical presence in the organizations' offices, speaking to staff and volunteers, meeting families in the community, and visiting local schools, I became a recognized face and news of my research soon began to spread. People were aware of the research, and this awareness aided their decision about whether to take part. Personally this also helped me to gain a better understanding of the social and cultural influences on the area. The added benefit that the organizations both had good reputations in the communities and were thought of highly also helped me to be seen as trustworthy and encouraged people to be willing to talk to me and see how they could participate.

That is not to say that recruitment was a simple process. In common with many countries in the Global South, Zambia has much higher proportions of younger people than middle-aged or older people. Thus in both urban and rural areas I effectively had no shortage of people from which to select my sample. However, those who identified as not having a significant caring responsibility were in greater number than those with a caring responsibility. In the urban area particularly, I was inundated with offers to participate in the research – something that on one occasion meant I had to adapt my research methods, as will be discussed later. It also brought about the question of how do you 'choose' one individual over another to take part in the study when one person's life story is just as valid as the next?

In contrast to this, recruiting young people who were caregiving took much longer than anticipated. The specific nature of the recruitment criteria for these young people meant they were much fewer in number. Differing understandings of illness and disability meant that young caregivers were sometimes harder to identify, while the stigma attached to many conditions may have increased reluctance to come forward (Evans and Atim 2011; Evans and Day 2011).

I had feared that community workers attached to the NGOs I worked with would try to highlight 'worst case' scenarios that they felt would illustrate the point better. However, I actually found that they sometimes tried to select families that didn't meet the criteria but which they thought would benefit: 'briefing' them beforehand, or even in my presence in their vernacular language, on the things they should say so that they could be involved. For example, having walked for some five miles in the bush to meet a young man who I was told was caring for his sick mother, I was actually introduced to a man well over 30 in age, married with children, but whose elderly mother lived with him as he had inherited the family land. It was obvious he did not fit the criteria, yet he still insisted he was 17! As this was a preliminary visit, I was able to say that I was only meeting families at that time and that I would be in touch later if he was the best person to take part. With that I politely excused myself and then tried to reiterate to the community workers exactly who it was I was looking for (see also Fagerholm in this volume, for whom village leaders chose research participants).

A trio of ethical considerations: informed consent, anonymity, and participant safety

One of the key ethical issues that I wanted to ensure was that all participants would take part in the research having given fully informed consent. 'Informed consent' has come to the forefront in recent years as the ethics involved in conducting research have become increasingly important (Brydon 2006). It is now considered essential that all research participants understand why the research is taking place, their role in it, and how the information they give will be used (Lindsay 2000). It is the responsibility of the researcher to provide clear, detailed, and factual information about the study, its methods, and how the person will be involved, and particularly that they are not obliged to participate and can withdraw from the study at any time (Masson 2000; Wassenaar 2006).

Accessible information leaflets were designed to introduce the research to young people outlining who I was, the purpose of the study, and what their participation would involve. I had designed something similar in my previous work in a children's charity and so already had a good idea of the level at which to pitch the leaflets, the language to use, and the design layout. However, I still consulted my supervisors about this and used my pilot month to trial them and make sure they were appropriate. During the main fieldwork, these were given to the young people in advance of the research taking place, either by myself or by the community workers from the organizations.

The leaflets were produced in English, as this is the national language of the country. I had originally planned to translate them into local languages as well if English was not spoken, but levels of literacy and confidence in reading were often low, even in local languages, and participants indicated their preference for having the same information offered to them verbally instead. Whether the participant spoke English or not, the physical act of being given the leaflet, which had a photograph and colorful writing on it, was well received. One young person's chronically ill father carried it around with him with great pride, almost as a status symbol because his family had been chosen to take part.

Consent to participate in the research was also negotiated verbally from young people in line with other research studies in Africa (Evans and Becker

2009, 2010). In many African communities, low levels of literacy coupled with suspicion at signing documents means that verbal consent is preferable. Young people in my study also saw low literacy levels as an embarrassment and something they did not like to admit. Offering the option of verbal consent overcame this problem, highlighting their agency and importance to the study regardless of their ability to read or write.

A set 'script' was prepared to be spoken to the participant at the start of each interview, which also sought consent for the digital recording of the interview and for taking a photograph of the young person and their home at the end of it. All the participants agreed to have the interview recorded, although not everyone was happy to have their photograph taken. Those who refused digital or photographic recording could still take part in the interview if they wished. Verbal consent was also obtained from parents, relatives, or guardians for young people under the age of 16 who wanted to participate in the research. As far as I am aware, both parents and young people understood the information and were given many opportunities to ask questions or for clarification if they wanted to. While the concept of 'research' itself may have been quite alien to them, they understood that I wanted to find out about their lives and that what they had to tell me was important.

Because of the sensitive nature of the research topic, the safety of the participants was the utmost priority at all times and I put in place a number of measures to minimize any risk. Respect for privacy, confidentiality, and rights to anonymity to protect young people's identities (Masson 2000) were paramount at every stage of the research, and the individuals themselves chose pseudonyms by which they would be identified. A small number of the participants wanted me to use their real names and indicated that they were happy for their stories to be told and recognized. However, for the majority their home situations were not something to be proud of in their eyes and they were happier that their accounts would be anonymized. Pseudonyms were therefore used for everyone.

It was made clear to the young people that they were not obliged to take part in the study and could withdraw at any time. Thankfully none of them did. Referral procedures were also discussed and clarified in case sensitive information should be disclosed in the research about the participants or others, and appropriate referral measures were agreed in advance and guided by the organizations (Masson 2000; Evans and Becker 2009). All of these issues were highlighted in the information leaflet and re-emphasized verbally at the start of each interview.

In some ways all of these ethical procedures felt quite bureaucratic when most people just wanted to go ahead and talk to me; they had already said 'yes' to taking part, so that was good enough in their eyes. To run through all the informed consent information verbally at the start of each interview took a great deal of time, especially when I knew that I had a substantial number of questions ahead to ask them. I would often find myself paraphrasing the points, or trying to run through them more quickly, as I could sense people's impatience to get started. Some were giving up valuable time in the fields or looking for work to speak to me, while others were due in school just a short time later so there was no time to lose. However, my previous research training in NGOs, as well as guidance from the university, meant I would not have been willing to forgo this. At least I knew my participants had all the information they needed and I had done everything in my ability to ensure their safety.

Applying ethical thinking to the methods

I also had to give ethical consideration to my research methods. As 'children's geographies' and the 'new social studies of childhood' (James *et al.* 1998) have become more established, research has sought to find 'ever more imaginative ways to allow children and young people to articulate their experiences and concerns in words, often encouraging them to employ other media of representation as stimuli' (Van Blerk *et al.* 2009: 3). I wanted to balance the need to gain information from the young people involved with methods that would interest them, while minimalizing the risk of distress.

I needed methods that would offer young people an opportunity to speak openly, share their views, and tell me about their lives in a way that would make them feel comfortable and integral to the research, but that would also respect when they felt they had said enough or did not want to pursue an issue further. I wanted not only to ask young people about their lives generally, but also to find out more specifically about illness and disability in their families, currently and in the past. Mindful that this could bring up topics such as HIV and AIDS (despite not asking directly), bereavement, poverty, relationships, cultural practices, and their socio-economic situation, I was aware the research would need to be both sensitive as well as probing.

During my pilot month in Zambia, young people indicated to me that they liked being given the opportunity to talk about their lives. Indeed Stephen and Squires (2003: 161) write that 'we must simply listen to what young people themselves have to say when making sense of their own lives.' I therefore decided to use individual semi-structured interviews which could be conducted on a one-to-one basis or with the help of an interpreter. These would enable me to gain an in-depth understanding of the young people's perspectives and experiences, guided partly by myself, but would also allow them to shape the conversation with their own ideas and thoughts (Willis 2006). This format would allow them to speak openly with me while still maintaining confidentiality and anonymity (Evans and Becker 2009). I would also be able to manage the interview situation and hopefully be able to recognize when they did not want to discuss an issue or felt they had said enough, and to move the questioning on.

However, recognizing that not all young people may be comfortable talking about their lives, I decided to complement the interviews with a life-mapping activity where young people developed a timeline of their lives to map 'self-defined "fateful moments" of transition' (Worth 2009: 1052) from their past and that might occur in the future. This design was based on Kesby *et al.*'s (2005) notion of 'participatory diagramming methods' and was inspired by the work of several researchers working with young people including Worth (2009), Van Blerk (2006), Kesby *et al.* (2005) and Young and Ansell (2003).

This method gave the young people an opportunity to provide a visual account of their lives, which they could complete independently or with the help of the interpreter if they wished. In the early stages of the research this was a particularly useful tool. It helped to clarify points from the interview and elicit more information as, once completed, it was also used as the basis of a discussion. This helped me to understand the young person's interpretation of the activity and the significant events that they had highlighted. Using such a method not only built on the interview process but often helped the young people to contextualize events that were complex to articulate or painful or difficult to talk about (Worth 2009).

The life-mapping activity was particularly well received by those who had a higher level of literacy but often had little opportunity to use their skills in their day-to-day lives. When I asked how he found the activity, one young man commented how much he had enjoyed it as it was a long time since he had written anything on paper and he enjoyed using writing. In this exercise, it was not uncommon for these young people to write more of a biography of their lives which they would then read to me. For young people with lower levels of literacy, my interpreter supported them in producing a timeline and we discussed it as it progressed.

I found that the timeline became less useful (or dare I say it, even a chore sometimes) as the research progressed. This was probably because I was becoming more practiced and familiar with the questions, so the interviews were longer and more detailed. In such cases the timeline sometimes became slightly repetitive of the interview and was perhaps viewed less positively by participants who were getting tired by that stage. It worked best with participants whom I could interview on one day and then complete the timeline with on a different occasion, usually a few days later. This gave them a break in between and also a chance to reflect on what they had discussed with me and potentially come up with other significant events that they may have forgotten on the day of the interview. Unfortunately, this was only possible with participants that were easy to access, usually those in the urban areas. For those further afield to whom I had to travel some distance to meet, or those who had other commitments, both activities often had to be completed on the same day, which could be draining for both the participant and myself.

While the data collection methods were planned and given ethical approval in advance of the research taking place, I needed to be able to be adaptable, and to do so ethically. For example, on one occasion it was necessary to change from conducting individual interviews to conducting focus groups when 12 young people who volunteered as peer educators were brought together by one of the organizations to participate in the research. I did not have the resources to interview them all individually, nor was it appropriate for the study. However, their views and experiences were still interesting and valuable and I did not want to disappoint them. Instead they agreed to take part in two focus groups of six young people each. The same ethical procedures took place in terms of informed consent, but I adapted the questions to become more about general awareness of caregiving and transitions to adulthood and less about personal circumstances so that they did not feel they had to disclose sensitive information in front of their peers.

Ethical dilemmas of recognizing young people's participation

I wanted to recognize the young people's participation in my research and thank them for giving up their time, sometimes quite a significant part of their day (see also Staddon in this volume for a detailed discussion on 'giving back'). The majority of the young people came from very low-income families and were often living in difficult circumstances. Whether they had caregiving responsibility or not, many had economic responsibility for their families, often as the sole income generator. I believed that it would have been unethical not to give them something in recognition of this, to compensate for any lost earnings, and also to recognize their value and contribution to the study.

In line with current social research practice (Evans and Becker 2009), each participant received a small financial payment as a gift for taking part. An appropriate amount was decided in consultation with local NGO workers acting as gatekeepers to the young people and their families, and each individual received the same amount. I also took with me to each interview a drink and a small snack for the participant and myself in recognition that not only would they be working with me for up to two hours, sometimes more, but that they might not have had breakfast that morning, or a full dinner the night before. I felt that a snack would be a nice thing to have while we took a break between research methods, often relaxing the atmosphere as we chatted informally while we ate, and would also help their (and my) concentration.

The best way to recognize the participation of individuals in research has been widely discussed in the literature (Brydon 2006; Van Blerk 2006), but the giving of cash payments is a contentious issue (Cree et al. 2002). In some cases it has been argued that payment is not appropriate and that it encourages the expectation that researchers will be a source of financial support. As highlighted by Evans and Becker (2009: 79),

> long-term academic goals of building knowledge about previously hidden, marginalized groups, which may help to improve the circumstances of others in future, are difficult to reconcile with the immediate practical needs for support that families involved in the study identif[y].

I did consider alternatives to payments, such as giving participants paper and pens (Cree et al. 2002) or buying something that everyone in the community could benefit from (Brydon 2006), but I decided that this was not appropriate in my study based on the individual needs and physical distances between the young people involved. The only exception was for the young people who took part in the two focus groups; I did not have the resources to pay them all individually, so I provided refreshments as a 'thank you' for their contribution. This included soda (the local term for carbonated drinks such as Coke or Fanta), cookies, and potato chips, which are snacks they would not normally be able to access or afford so were perceived as a rare 'treat.' They appeared to appreciate this just as much as they would have being paid, and it actually made each focus group a more relaxed affair with an informal, fun atmosphere.

Much of the discourse on payments also focuses on the appropriateness of giving money to 'children' as opposed to older young people or youth. Concern about how they will spend the money, as well as the fact that it may influence their relationships with their peers (Van Blerk 2006), means alternatives are advocated. However, I believe this fails to recognize the levels of responsibility that these young people already hold, many from a much younger age. Offering payment respected the young people's agency and competence to receive money and spend it appropriately (Evans and Becker 2009).

To try to counter any expectation of further payment or financial support in the future I included in the information leaflet, and also gave verbally before each interview took place, information clarifying my position as the researcher offering a one-time payment only and that I did not have the means to offer them further support in the future. Despite this, a small number of people did still ask for help, particularly in terms of school sponsorship. While they seemed to accept my explanation that I was unable to help them further, I felt that they thought I was lying; in their eyes I was still a rich, white Westerner with access to resources. Even if I did not have money myself, many thought that I would have links to other bigger agencies who would come in with food, money, or housing and relieve their situation. This is hardly surprising in a country where aid agencies seem to be everywhere and every other car on the road has a UN or NGO logo on the side. I also had to consider these expectations in terms of whether they might influence the answers participants gave if they thought this might improve their chances of getting access to future help and support (Van Blerk 2006).

A rewarding experience

This chapter has provided a reflexive account of how ethical considerations played a significant role in my doctoral research conducted with young people in Zambia. In fact, they underpinned all aspects of my interaction including identifying the participants, ensuring their consent to take part and their safety in the research process, choosing the most appropriate data collection methods, and decisions about how to recognize their participation. In addition to this, by having to think about and formally present my ethical procedures in advance, I felt that I was more prepared for actually undertaking the fieldwork, had more confidence in the field that I was acting appropriately, and had the ability to be adaptable while remaining within an ethical framework.

Conducting research with young people is rewarding, interesting, and often fun but it is important to remember that even older young people, or youth, can still be vulnerable and need to be protected (see also the chapters by Brooks, Skinner, and Taylor in this volume on conducting research with people who are vulnerable in other ways). In a cross-cultural setting it is particularly important to acknowledge where power relations between those doing the research and those being researched are further heightened by socio-economic and cultural divides. However, respect should also be given to young people's agency and their ability to make informed decisions, and recognition given to the positions they hold within both their families and communities. Offering payment to compensate the time they give to a research study is one way of doing this, but giving them the opportunity to contribute to research and have their voices heard in a safe and confidential setting is far more valuable.

It may seem from reading this chapter that working ethically with young people requires a lot of effort. Certainly there is a lot to consider before entering the field, both in terms of good practice on your own behalf and also to satisfy the requirements of the institution or organization with which you are working. However, to avoid working with young people based on perceptions of effort would be to miss out on hearing the views of a large and diverse proportion of society. In Zambia alone, young people under the age of 25 make up approximately 66 percent of the population (Central Statistical Office 2012). To exclude these voices would exclude the majority of the country and fail to recognize their agency as the best people to comment on their own lives.

My previous experience of conducting research with NGOs supporting children and young people in the UK meant that all this preparation was not alien to me; I had always been trained to do research in an ethical manner, so it did not feel as though I was going over and above the expectations of a research project. In previous research projects I had always taken a similar route of designing information leaflets, preparing interview and focus group schedules, and ensuring I gained informed consent; to me it would actually have felt wrong to not do this. It may not be obvious when working with young people that they understand the reasoning behind ethical procedures or how the study is designed with ethics in mind. However, I know that considering these issues helps to ensure that I implement an appropriate research study and that, as much as is possible, I have considered the best interests of the participants, the organizations I want to work with, myself, and my research institutions. If seen as part of the process and implemented from the start, research ethics simply become part and parcel of the rewarding experience that is working with young people.

Recommended reading

Alderson, P. and Morrow, V. (2004) *Ethics, social research and consulting with children and young people*, Essex: Barnardo's.

This book is short but accessible and easy to read. It is split into three logical sections – 'The planning stages,' 'The data collecting stages,' and 'The reporting and follow-up stages' – which guide you through designing a research project from start to finish. Ethics is a key element of the book, including working with ethics committees and the consent and payment of young people.

Desai, V. and Potter, R. (eds) (2006) *Doing development research*, London: Sage.

This book provides an excellent starting point for anyone (particularly those with little or no experience) about to undertake fieldwork or research overseas. All the chapters provide interesting advice, hints, and tips, but in terms of my own research I found two chapters particularly useful. Chapter 3 on 'ethical practices in doing development research' by Lynne Brydon offers advice concerning informed consent and issues of power, as well as providing an excellent overview of things to think about before, during, and after undertaking research. Chapter 6 on 'Working with children in development' by Lorraine Van Blerk provides a thoughtful and detailed account of applying ethical thinking when working with particularly vulnerable children and young people, based on Van Blerk's experiences of researching street children in Uganda.

Evans, R. and Becker, S. (2009) *Children caring for parents with HIV and AIDS: Global issues and policy responses*, Bristol: Policy Press.

This book provides a comprehensive account of Evans and Becker's experiences of working with young carers in both the UK and Tanzania. This book was useful to me because of the specific nature of the research group and its links with my own research participants. However, Chapter 3 is also particularly relevant to the consideration of ethics and describes in detail the research process, particularly the involvement of children and young people in research, the positionality of the researchers undertaking the fieldwork, the methodology chosen, and the ethical considerations negotiated, including informed consent and anonymity.

References

Alderson, P. (1995) *Listening to children: Ethics and social research*, Ilford, Essex: Barnardo's.
Alderson, P. and Morrow, V. (2004) *Ethics, social research and consulting with children and young people*, Ilford, Essex: Barnardo's.
Ansell, N. (2005) *Children, youth and development*, London: Routledge.
Barker, J. and Weller, S. (2003) 'Geography of methodological issues in research with children,' *Qualitative Research* 3 (2): 207–227.
Brydon, L. (2006) 'Ethical practices in doing development research', in V. Desai and R. Potter (eds), *Doing development research*, London: Sage.
Central Statistical Office (2012) *The Monthly*, vol. 110, May, Lusaka: Central Statistical Office.
Clacherty, G. and Donald, D. (2007) 'Child participation in research: Reflections on ethical challenges in the southern African context,' *African Journal of AIDS Research*, 6 (2): 147–156.

Cree, V. E., Kay, H. and Tisdall, K. (2002) 'Research with children: Sharing the dilemmas,' *Child and Family Social Work*, 7: 47–56.

Duckett, P., Sixsmith, J. and Kagan, C. (2008) 'Researching pupil wellbeing in UK secondary schools: Community psychology and the politics of research,' *Childhood*, 15: 89–106.

Evans, B. (2008) 'Geographies of youth/young people,' *Geography Compass*, 2 (5): 1659–1680.

Evans, R. (2011) '"We are managing our own lives…" Life transitions and care in sibling-headed households affected by AIDS in Tanzania and Uganda,' *Area*, 43 (4): 384–396.

Evans, R. and Atim, A. (2011) 'Care, disability and HIV in Africa: Diverging or interconnected concepts and practices?,' *Third World Quarterly*, 32 (8): 1437–1454.

Evans, R. and Becker, S. (2009) *Children caring for parents with HIV and AIDS: Global issues and policy responses*, Bristol: Policy Press.

Evans, R. and Day, C. (2011) *Inheritance, poverty and HIV/AIDS: Experiences of widows and orphaned youth heading households in Tanzania and Uganda*, Chronic Poverty Research Centre Working Paper No. 185.

Greene, S. and Hill, M. (2005) 'Researching children's experience: Methods and methodological issues,' in S. Greene and D. Hogan (eds), *Researching children's experience: Approaches and methods*, London: Sage.

Holloway, S. L. and Valentine, G. (2000) *Children's geographies: Playing, living, learning*, London: Routledge.

James A., Jenks C. and Prout, A. (1998) *Theorizing childhood*, Cambridge: Polity.

Kesby, M., Kindon, S. and Pain, R. (2005) '"Participatory" approaches and diagramming techniques,' in R. Flowerdew and D. Martin (eds), *Methods in human geography: A guide for students doing a research project*, Harlow: Prentice Hall.

Lindsay, G. (2000) 'Researching children's perspectives: Ethical issues,' in A. Lewis and G. Lindsay (eds), *Researching children's perspectives*, Buckinghamd and Philadelphia: Open University Press.

Masson, J. (2000) 'Researching children's perspectives: Legal issues,' in A. Lewis and G. Lindsay (eds), *Researching children's perspectives*, Buckingham and Philadelphia: Open University Press.

Morrow, V. and Richards, M. (1996) 'The ethics of social research with children: An overview,' *Children and Society*, 10: 90–105.

Robson, E., Porter, G. and Hampshire, K. (2009) '"Doing it right?" Working with young researchers in Malawi to investigate children, transport and mobility,' *Children's Geographies*, 7 (4): 467–480.

Scott, J. (2000) 'Children as respondents: The challenge for quantitative methods,' in P. Christensen and A. James (eds), *Research with children: Perspectives and practices*, London: Falmer.

Stephen, D. and Squires, P. (2003) '"Adults don't realise how sheltered they are": A contribution to the debate on youth transitions from some voices on the margins,' *Journal of Youth Studies*, 6 (2): 145–164.

Valentine, G. (2003) 'Boundary crossings: Transitions from childhood to adulthood,' *Children's Geographies*, 1 (1): 37–52.

Van Blerk, L. (2006) 'Working with children in development,' in V. Desai and R. Potter (eds), *Doing development research*, London: Sage.

Van Blerk, L., Barker, J., Ansell, N., Smith, F. and Kesby, M. (2009) 'Researching children's geographies,' in L. Van Blerk and M. Kesby (eds), *Doing children's geographies: Methodological issues in research with young people*, London and New York: Routledge.

Wassenaar, D. (2006) 'Ethical issues in social science research', in M. T. Blanche, K. Durrheim and D. Painter (eds), *Research in practice: Applied methods for the social sciences*, 2nd edn, Cape Town: University of Cape Town Press.

Willis, K. (2006) 'Interviewing,' in V. Desai and R. Potter (eds), *Doing development research*, London: Sage.

Worth, N. (2009) 'Understanding youth transition as "becoming": Identity, time and futurity,' *Geoforum*, 40: 1050–1060.

Young, L. and Ansell, N. (2003) 'Fluid households, complex families: The impacts of children's migration as a response to HIV/AIDS in Southern Africa,' *Professional Geographer*, 55 (4): 464–476.

Young, L. and Barrett, H. (2001) 'Adapting visual methods: Action research with Kampala street children,' *Area*, 33 (2): 141–152.

18 Power play: ethical dilemmas of dealing with local officials and politicians

Experiences from post-tsunami research in Sri Lanka

Kamakshi Perera-Mubarak

> There is also a request I have to make of you, but please don't misunderstand me. I told you these details not with expectation of any favor. But, this job of ours is run with great difficulty. On the 10th we have the elections coming up – we don't get a cent for our activities! So, I would like to ask you, if you are able to in any way, to give us some support – a contribution.

This request came in an interview with a politician and former Minister in Sri Lanka. The purpose of the interview was to find out how he perceived the roles, activities, and behavior of households, aid officials, and other politicians in the area in the aftermath of the 2004 Asian tsunami. Yet the timing of the interview coincided with impending Southern Provincial Council Elections.

Researchers are not always able to walk away from interviews merely with recordings and notes. Through their presence in the field, relationships they form with the research community, rapport they build with respondents, and methods of enquiry they employ, researchers create bonds with the researched. These bonds can result in difficult circumstances for the researcher, giving rise to ethical dilemmas about appropriate behavior. The ideal of the researcher's impartiality is not realistic in most qualitative fieldwork. In the above case I did not want to give an immediate negative response that might have offended or disappointed my interviewee. His request came toward the end of the interview, but I was still conducting research in the area and did not want him to endanger forthcoming interviews. I was put in an awkward position with implications of unethical dealings on my part. I will reveal my response later in this chapter after exploring the difficulties related to behaving 'ethically.'

Much research in the field of development is focused on poor and disadvantaged members of society, but some research involves individuals who hold or have held privileged positions in society and who therefore are in possession of more wealth, political power, or influence than the general public. Such research on the powerful is known as 'studying up' and involves a different set of practical challenges and ethical complexities compared to studies where the researcher tends to have more power than the researched.

'Studying up' can be problematic as respondents have the power to create barriers, shield themselves from scrutiny, and resist the intrusiveness of social research (Duke 2002). The researcher can face ethical dilemmas in terms of gaining access, responding to corrupt practices, effective positioning, ensuring confidentiality of data and anonymity of respondents, and upholding reciprocal relations. In this chapter I relay my experiences of dealing with local officials and politicians (see the chapters by Chopra, Dam and Lunn, and Wang in this volume for examples of dealing with other types of elites).

The purpose of my doctoral fieldwork in Sri Lanka was to explore how livelihoods were recovering after the 2004 tsunami, with specific reference to two fisher villages in the Hambantota District in the Southern Province (Mubarak 2011). I probed commonly overlooked social and political processes in post-tsunami livelihoods recovery using a qualitative ethnographic methodology that examined narratives emerging from households, local officials of government organizations and NGOs, office bearers of community-based organizations, local politicians, village leaders, and key informants. This included an investigation into practices of corruption and political interference at the village level. The aim was to advance conceptualizations of informal politics in South Asia. Local government officials played a key role in the official distribution of aid in these villages, while politicians were significant as an informal means of accessing aid. What follows are my reflections on interviewing these powerful people, focusing on five key interrelated ethical issues: access, corruption, positionality, confidentiality, and reciprocity.

'Getting in' and 'staying in'

Having connections and being 'street smart' is the key to gaining access to and cooperation from powerful people. In the immediate aftermath of the 2004 tsunami, I worked as an intern and volunteer in the organization and distribution of post-tsunami relief conducted by various government agencies, NGOs, and private donors. Following this, I worked as a volunteer over two periods in projects identifying livelihood needs run by two NGOs in parts of Southwest and South Sri Lanka. My observations during these stints became points of departure for fieldwork conducted as part of an Honors degree thesis and a later research paper on post-tsunami livelihoods recovery (Mubarak 2007). The local knowledge that I gained through these early experiences helped to facilitate an easier access to local officials and politicians during my doctoral research.

The network of personal contacts I made along the way also helped me to identify and access many officials and politicians with whom I had no prior links. What Walford (1994) calls 'personal sponsorship' of the study was crucial in the case of politicians who, without an introduction through an individual known to them, were very unlikely to grant access (see also Wang in this volume on the use of personal social networks to identity potential inter-viewees). Interviews invariably had to be preceded by an introduction made by a contact through a telephone call. The interview itself would be started off by mentioning this contact. This pleased interviewees since it was made clear that they were recognized within their relevant networks. Moreover, it helped establish my legitimacy and reduce perceived threat. As Cassell (1988: 95) writes, the researcher of the powerful needs many of the characteristics of the social climber: 'everyone who might possibly know someone must be con-tacted and asked if they will give introductions, vouch for one, and otherwise help one's enterprise.'

Political happenings in the country during the time of my fieldwork mat-tered greatly in terms of access to politicians. Impending Southern Provincial Council Elections had had mixed impacts on my research conducted in the previous three months. In the midst of election fever, local political candidates were eager to be seen as cooperative and were not likely to refuse to be inter-viewed by a student conducting research in their local area. But securing an appointment for an interview was made very difficult due to the politicians' heavy schedules and the large numbers of local people who also flocked to their offices, making use of the occasion to seek favors. For example, despite an introduction secured through a contact, an interview with a political can-didate was made possible only by my waiting in line outside his house with throngs of others from 6:00 a.m. until 1:00 p.m. When I finally reached the front of the line, the politician concerned was unable to spare time for a meeting (see also the chapters by Chopra and by Dam and Lunn in this volume for similar frustrations).

In tight-knit communities, associations with politicians can be perceived with suspicion and can imply a political affiliation. I had to give careful consideration to the timing of interviews with politicians in relation to those with households. Conducting both sets of interviews in the same timeframe might have jeopard-ized my attempts to access and establish a rapport with household respondents. Thus I interviewed politicians during my second stint of fieldwork, after most household interviews had been concluded.

Such strategic timing and cautious behavior on my part and the personal sponsorship of my research did not eliminate the ability of officials to create obstructions. A local government official, in office during the time of the tsunami, was introduced to me by a local politician. During a brief meeting with them both, a date and time was set for me to conduct a one-on-one interview with the official. However, at the appointed time and place, the official could not be reached. When contacted on the telephone she claimed to be out of station, although it was revealed later that she was actually in the building. A second appointment that was scheduled was similarly abandoned. It was clear that she did not want to be interviewed.

The official in question had been accused by many of my interviewees of having engaged in corrupt practices such as granting special favors to friends and relatives and putting people on beneficiary lists in exchange for bribes.

Anecdotal evidence suggested that she had been interdicted from her position and was facing a court case at the time. Reiterating the experiences of other researchers (Johnson 1992; Cook 1993), I may have been seen as having power to open out the life of this official for contestation or blame by other groups and the general public, and hence viewed as a threat.

I could have probed the case further by securing a meeting with the official through the intervention of the politician through whom the original contact was made, but I refrained from doing so for two reasons. Fisher villages comprise closely integrated communities, and any attempt on my part to dig deeper into the incident to determine the reason for the official's reluctance to be interviewed might have drawn unnecessary attention to my work, impeding further research in the village. It could also have seriously affected the official's future. Being sensitive to these ethical and practical considerations was paramount during fieldwork.

Tackling corruption

A key objective of my research was to examine the practices and perceptions of corruption in the study villages (see also Brooks in this volume, on researching corrupt practices). Directly probing aspects viewed as 'immoral' or 'illegal' in the highly charged and politicized environment of post-tsunami recovery would have been counterproductive, since it would threaten betrayal of the trust or confidence of officials and politicians with whom I had developed a strong rapport. My strategy was therefore to listen to people's 'stories.' I waited until at least halfway through the discussion, when I felt that I had established a rapport with my interviewee, before asking contentious, critical, or tricky questions about corruption. Plunging into these types of questions at the outset might have alienated the official or politician, who might have then become unforthcoming or defensive.

Deciding how to react when I knew my respondents had engaged in corrupt practices was not an easy task. After the tsunami, the loss or destruction of fishing boats and gear had to be substantiated by an entitlement certificate detailing the type and amount of loss, a methodology introduced by the Ministry of Fisheries and Aquatic Resources to help government and non-government aid actors identify genuine beneficiaries. A former Fisheries Inspector was accused by many villagers of issuing fraudulent entitlement certificates in exchange for bribes. During an interview with this official, I enquired about irregularities under his authority. He responded with the following:

> If we say there was no bribery at all, these things in the newspapers would have to be lies, wouldn't it? People are human and officials are human so there can be situations where they may try to help in various ways.

It was clearly an evasive answer. Yet I chose not to probe the case further as I did not want to give him the impression that I was searching for confessions.

I faced another dilemma when interviewing the president of a village Fisheries Cooperative Society (FCS). The officer had been a member of this FCS for over ten years and had been the president of both the FCS and the larger district FCS Association for a number of years. Using an international NGO donation acquired via the district FCS Association, the officer had launched a credit program offering interest-free loans to households which had members who died in the tsunami. Most villagers were of the view that he had manipulated his position of authority to secure FCS funds for his personal benefit and for his relatives and close associates. My challenge was how best to probe him about the accusations of irregularities in his behavior without upsetting him while still gathering useful responses for my study.

Monthly general meetings of FCSs should be attended by a government representative from the District Fisheries Office (usually the Cooperative Officer or Fisheries Inspector) to monitor decision-making and management. When I enquired from the officer about this aspect of the FCS, his reply was rather aggressive:

> I have studied up to the Advance Level Examination and held the post of President in societies since 1993. I have participated in many workshops and have extensive knowledge of the fisheries sector. I have held several posts in organizations related to this field. So, with that experience and knowledge I don't need an officer to sit in on meetings. Besides, it is because of people like us that the cooperative movement developed.... So, I know the cooperative law – I know what is meant by 'cooperative fisheries' and the rights of the members of a FCS. How a democratic fisheries cooperative should function, the responsibilities of the officers, the rights of the members – all that I know. Not all, but I know enough to make decisions. Whatever problem that arises at a general meeting, I have the capacity to resolve it. I don't need an officer. There are no problems that arise which can't be resolved by the President, by me. If I think it necessary, I will invite them.

The statement was an accurate reflection of his position of power. I felt rather intimidated; I thought my question had been too blunt and confrontational. In retrospect, however, I think that too great a concern with rapport on my part may have led to a bland answer being insufficiently challenged. It was important to ensure that my research agenda was followed rather than my respondent's.

'Insider' or 'outsider'?

My identity and positionality played a critical role in influencing the study. Ozga and Gewirtz (1994) suggest that access is more likely to be granted if the interviewer seems 'perfectly harmless.' My position as a young student meant that I was associated with 'naivety' and 'willingness to learn.' In some cases, being a student from an overseas university worked in my favor: politicians and officials considered it flattering that someone from an academic institution in the West (an 'outsider') should be interested in their opinions. In other cases, highlighting my Sri Lankan student identity and downplaying my English-speaking, middle-class social background (an 'insider') was more appropriate in order to stimulate flowing responses from the interviewee. Cultivating an ability to represent myself according to the situation was of critical importance. Being female may also have helped me to present a non-threatening image (Easterday *et al.* 1977; Klatch 1988). In a sexist Sri Lankan society, most of the powerful positions are still held by males. Thus young female researchers are at an advantage in being perceived as 'harmless,' non-threatening, and without power (see also the chapters by Dam and Lunn, Godbole, and Kovàcs and Bose in this volume on the strategic use of aspects of researcher identity).

Recognizing ethnic lines of identification was also vital in facilitating rapport between me and my respondents and thereby negotiating access to views and opinions beyond the 'official line.' Ethnicity is a major dividing line in Sri Lankan society. The Hambantota District comprises predominantly Sinhalese, Muslim, and Sri Lankan Malay groups. Thus my mixed ethnicity as an Indian Muslim and a Southern Sinhala-Buddhist helped me to gain an 'insider' status, being treated in solidarity among both Muslim and Sinhala-Buddhist respondents.

A critical challenge was the extent to which I ought to make clear my own views. I tried to present myself as 'an intelligent, sympathetic, and non-judgmental listener' (Cassell 1988: 85), neutral toward personal issues in post-tsunami recovery yet supportive of people's desire for an effective post-tsunami recovery. This was not always a far-sighted approach. Presenting a 'neutral' character at times conveyed an impression that I was unfeeling and therefore undeserving of the respondent's attention and time. It may have also raised an ethical issue: the concealment of my true identity from respondents.

I did on occasion express personal beliefs and values and take particular stands during interviews; for example, during an interview with a politician I denounced the mismanagement of aid by a particular local government official. Giving such a formal indication of my opinion may have biased the outcome of the interview. I justify this action as part of an essential undoing of an otherwise exploitative, hierarchical, and unproductive research encounter.

Respondents themselves are subjects with agency, history, and their own idiosyncratic command of a story. Many householders, politicians, and officials warned me about twisted facts on post-tsunami aid distribution that might be put forward by other politicians and officials in the village and offered to tell me the 'real story.' Some continued to view me as having a political bias despite my having clearly explained the objectives of my study and assured them of the anonymity of any information they might give. Others appeared somewhat guarded in their responses or appeared to say the 'right thing' fearing that the information might be fed back to other politicians and officials. In fact, the power that politicians could wield in appointments, transfers, and promotions of public servants was such that the latter tended to be vague in their responses on political intervention and almost never responded unfavorably (Perera-Mubarak 2012).

Assuring confidentiality and anonymity

Protecting the anonymity of research participants is a fundamental ethical principle in qualitative research. As opposed to household respondents, who were chosen to be representative of a tsunami-affected population, politicians and officials in my study were selected specifically because of who they were and the positions they held. Many were public figures such as ministers, members of parliament, presidential advisors, and directors of key government institutions. Offering them anonymity was difficult as it was not only 'what' was said that mattered but also 'who' said it (see also Chopra in this volume).

Practical matters such as the need to retain good relations for future interviews and research in the area were important for me. There was a possibility that disclosing information that was 'hidden' and critical of my respondent could impede my access to that research site again. It could also adversely affect future researchers in the area as they too may be excluded. Thus I made an attempt to reveal the information collected in the field in a manner that did not result in any negative outcome for respondents. I took extreme care in what I wrote where there was any doubt about interpretation. I used pseudonyms to ensure privacy and confidentiality. Additionally, I took steps to exclude certain identifying information on interviewed politicians and officials in the final thesis and related publications.

Protecting the anonymity of research participants while out in the field was not always feasible. Some respondents were particularly interested in knowing who else I had spoken to. At the start of my fieldwork, when I was asked this question I tried to give vague answers because I felt that any information of this nature would imply a breach in my assurances of confidentiality and anonymity. In the latter part of my fieldwork I realized that in such tight-knit communities respondents often already knew whom I had interviewed. Some had even discussed my presence in the village, research topic, and questions posed with previously-interviewed respondents. This raises the issue as to whether research participants themselves have responsibilities regarding anonymity and confidentiality, in addition to the responsibilities of the researcher.

Giving back

Many researchers face ethical dilemmas when 'giving back' to interviewees or research participants (see Staddon in this volume for a full discussion). I too encountered many difficult choices in interviewing elites. Gift-giving can be politically difficult when those who participated in the research benefit over non-participants in the same community. While I gave gifts to research assistants and contacts in appreciation of the specific help they provided during fieldwork, I needed an alternative approach for showing appreciation to research participants that was not perceived in the wrong way and did not set dangerous precedents. Very clearly communicating to politicians and officials that I valued their information and knowledge was a way of ensuring the research experience was rewarding for them. I also made an effort to interact socially with some respondents in between interviews so that they would feel valued and friendships could be cemented.

Yet giving back was not always a straightforward process and the example at the beginning of this chapter provides an apt example. The same politician followed up that request with another:

> My wife's sister's husband is our party organizer in England. What I would like to do is to have you meet with him and get an understanding of our work. If you would give me your personal address and contact number, I will make the connection. They will get in touch with you.

Outright refusal to satisfy the requests to make a donation and to meet relatives might have offended or disappointed the politician and carried the risk that he might jeopardize other interviews that I had lined up in the area. However, agreeing to satisfy the requests would have meant my getting involved in unethical dealings. Therefore, my response was that I was not sure how much I could commit at the time. I enquired about bank account details so I could make a preferred transaction on a later date. The interviewee said that he did not have these details and would get back to me if I gave him my local telephone number. I gave this number along with my contact details in England, thus partly satisfying his other request. To date, I have not had follow-up enquiries on either of these requests. Had I received a request for a donation or been approached by his relatives in England, I would have very subtly refused both by explaining that I did not want to in any way involve myself in politics but rather that my inquiry was solely for research purposes, with his interview time and thoughts contributing toward development programs and policies in Sri Lanka.

The above incidents emphasize the fact that the researcher runs the risk of behaving unethically when giving back, particularly to elite interviewees. Moreover, they reveal that power relationships during fieldwork are not fixed or unidirectional (see also Wang in this volume on ambiguous and shifting power relationships). The fact that a politician requested various contributions to the party for his services shows how researchers themselves can in some ways be overpowered by their subjects.

Playing the game

There is a tension between the ease with which we talk and write about ethical procedures and the realities of following these through in practice. I encountered various challenges in establishing, maintaining, and revising my ethical stance during research involving local government officials and politicians in post-tsunami Sri Lanka. Overcoming such challenges meant drawing on lessons learned in existing scholarship as well as capitalizing on 'street sense' and in-field negotiations.

Powerful people are usually less accessible, as well as being more conscious of their own importance, so problems of access are particularly significant. Exploiting pre-existing networks with those in power and using them as research 'sponsors' is an obvious way of easing access. However, researchers should take care in selecting such 'sponsors' since recommendations by the 'wrong' people – for instance, those perceived as being against a particular official or from an opposing political party – can backfire and hinder access.

My presence in the field represented multiple identities which shaped directly my relations and interactions with officials and politicians and what they were willing to divulge. Recognizing such lines of identification and the ways in which interviewees tend to 'place' the researcher is vital in facilitating rapport between interviewer and interviewee and thereby producing a rich, detailed conversation based on empathy, mutual respect, and understanding. A researcher should aim to adopt a role or identity that harmonizes with the values and behavior of the powerful without seriously compromising their own values and behavior and without inventing a false identity.

Sensitive topics such as corruption must be approached without destroying the relationship of trust between researcher and respondent and without giving the impression of searching for confessions. Maintaining confidentiality and anonymity, both in and beyond the field, is crucial as there can be real and perceived threats of libel. Negotiating ethical means of ending fieldwork by showing appreciation to respondents without setting dangerous precedents is also of paramount importance.

In sum, 'getting in' and 'staying in' are as important as playing the game through strategic approaches to investigating sensitive topics, assurances of confidentiality, and demonstrations of reciprocity. Interviews with local officials and politicians were perhaps more testing than those with households because of this 'game-like' nature. Yet the experience was fruitful and worthwhile. Corruption and political intervention in post-tsunami recovery were not topics I set out to investigate at the outset but rather intriguing aspects that I stumbled upon during fieldwork. It was the mundane and subtle expressions of local officials and politicians regarding the allocation of post-tsunami aid, observed during early interviews, which stimulated my exploration into practices and perceptions of corruption in post-tsunami Sri Lanka (Perera-Mubarak 2012).

Whether interviewing powerful or less powerful people, the field of study is a seriously relational experience, an 'interworld' (Crossley 1996) between the researcher and respondent where the subjectivities of both become entangled. In the field, my respondents and I consciously and unconsciously constructed our own meaning, objectified 'others,' recognized ourselves in them, and played on each other's performances accordingly to engage as best as we could. I think that the ability to understand such subtle relationships is the key to interpreting the responses of those in power. Therefore my experiences of dealing with powerful people during fieldwork have also been meaningful in my present-day work life, in interactions with high-level government and non-government officials.

Recommended reading

Crang, M. and Cook, I. (2007) *Doing ethnographies*, London: Sage.

This is an easy-to-read introductory and applied guide to qualitative ethnographic methods. It is a useful resource for students from a wide range of disciplines starting out on ethnographic research projects, especially in terms of its practical lens on how to accustom oneself to the environment and nuances of the study locale.

Duke, K. (2002) 'Getting beyond the "official line": Reflections on dilemmas of access, knowledge and power in researching policy networks,' *Journal of Social Policy*, 31, 39–59.

This paper provides a methodological and reflexive account of the key processes and issues encountered when researching policy networks involved in the development of prison drugs policy. It reveals the unique set of dilemmas and complexities that the researcher faces when studying powerful individuals within policy networks.

References

Cassell, J. (1988) 'The relationship of observer to observed when studying up,' *Studies in Qualitative Methodology*, 1: 89–108.

Cook, I. (1993) 'Constructing the "exotic": The case of tropical fruit,' paper presented at the Institute of British Geographers Annual Conference, Royal Holloway and Bedford New College, January 1993.

Crossley, N. (1996) *Intersubjectivity: The fabric of social becoming*, London: Sage.

Duke, K. (2002) 'Getting beyond the "official line": Reflections on dilemmas of access, knowledge and power in researching policy networks,' *Journal of Social Policy*, 31: 39–59.

Easterday, L., Papademas, D., Shore, L. and Valentine, C. (1977) 'The making of a female researcher: Role problems in fieldwork,' *Urban Life*, 6 (3): 333–348.

Johnson, M. (1992) 'A silent conspiracy? Some ethical issues of participant observation in nursing research,' *International Journal of Nursing Studies*, 29 (2): 213–23.

Klatch, R. E. (1988) 'The methodological problems of studying a politically resistant community,' *Studies in Qualitative Methodology*, 1: 73–88.

Mubarak, K. (2007) 'Analyzing post-tsunami livelihoods recovery: The case of masons in Polhena village, Sri Lanka,' Melbourne: School of Social and Environmental Enquiry, University of Melbourne.

Mubarak, K. (2011) *Everyday networks, politics, and inequalities in post-tsunami recovery: Fisher livelihoods in South Sri Lanka*, unpublished thesis, University of Oxford.

Ozga, J. and Gewirtz, S. (1994) 'Sex, lies and audiotape: Interviewing the education policy elite,' in D. Halpin and B. Troyna (eds), *Researching education policy: Ethical and methodological issues*, London: Falmer Press.

Perera-Mubarak, K. (2012) 'Reading "stories" of corruption: Practices and perceptions of everyday corruption in post-tsunami Sri Lanka,' *Political Geography*, 31: 368–378.

Walford, G. (1994) 'Reflections on researching the powerful,' in G. Walford (ed.) *Researching the powerful in education*, London: UCL Press.

19 Exercising my rights: ethical choices and moral predicaments in accessing government documents

Experiences from dealing with civil servants in India

Deepta Chopra

I've been waiting 45 minutes already … hope he does not ask me to come another day like he did last time … concentrate, look professional, look confident … oh finally … hello … oh sorry, he's on the phone. What are all these other people doing here? This is meant to be a confidential interview I thought … he is looking at his watch already … another phone call! Right, rephrase that … another phone call! He has to go … but that was only five minutes … perhaps I'll ask him the controversial question … ah smart move Deepta, that made him stop and talk to you for another two minutes … ok, I'll come back another day … but when? He has not given me a time … oh no … what a disaster.… Phew, that was hard, he told me almost nothing, and I can't quote the interesting bits because he said I was not allowed to.… Hello Mr Peon,[1] sure I'll come again tomorrow at 8 a.m. … when he will not even be here.

I was born and brought up in New Delhi in an upper-middle-class household, fluent in both English and Hindi. When I set out to interview 'elites' in New Delhi for my doctoral research, I was not too nervous despite the fact that I had never done research with elites before. I was, after all, from a similar background (Razavi 1993) and, having worked in the development sector in rural and urban sectors through international NGOs, was aware (or so I thought) of power dynamics being played by government officials. *I was an insider.* I had studied intensely the methodology of elite research (Herod 1999; Elwood and Martin 2000) and 'studying up' (Nader 1974), which is where this research was located. I was well aware of the different methodological considerations of interviewing elites (Hert and Imber 1995), their power (Richards 1996), and the fact that I, the scholar, was supplicant to the interviewee (McDowell 1998).

Thus, fairly confident in my insider status and familiar with the literature, I embarked on my research into the making of the Mahatma Gandhi National Rural Employment Guarantee Act (MGNREGA). The name of the University of Cambridge was useful in making initial contacts and I didn't hesitate to 'drop names' to ensure that I could get past the stream of personal administrators who guarded phone lines and cordoned off undesirable visitors from the senior officials' purview (see also Dam and Lunn in this volume on contrasting experiences of accessing

NGO and government elites in India). I managed to secure an interview with the civil servant who mattered at the Ministry of Rural Development. Never mind that it was after two months of insistent calling. Never mind that he had given me only half an hour at 5:30 p.m. at the end of the working day.

The first time that I met him (as recounted at the head of this chapter), I was kept waiting for 45 minutes, during which time I conversed with the peon about the differences between villages and big cities. When I was shown in, the civil servant was talking on the phone with someone, while there were three other people sitting with laptops in his room. I was interrupted no less than five times in the 10 minutes that I was given (instead of the promised 30 minutes) (see also Wang in this volume on dealing with telephone interruptions during interviews). He did not listen to my questions, except the last one which I had asked on purpose as a controversial probe. *Thank you, Bernard (2002)!* I was also told to not quote anything that he said. After that, as I was being rapidly shooed out of the office by the now-friendly peon, he said, 'Come at 8 a.m. tomorrow if you want to talk about this more.'

Sure, I thought, tears of frustration brimming in my eyes, *I'll come at 8:00 a.m. again, as if he is going to be in.* We all knew that no *sarkari babus* (government officials) were ever in their offices before 10:30 a.m., even though the official start time was 9:00 or 9:30 a.m. But, hey presto, when I arrived at precisely 8:00 a.m. I was shown into the office by the friendly peon, who remembered me from the previous day and did not press for any security clearance. Inside, the official sat in silence *on his throne* working on a document. Even the personal assistant was not in, and the phone was not ringing. He looked up from his document and said,

Ah, you decided to brave it again today. Now before you start asking me questions, let me tell you that you can record this interview if you want, but it is only for your aide memoire. You will not quote me or use any of this information that I am telling you. I am fully capable of writing all of this down myself and have no need to tell you anything. And if I ever hear that you have quoted me, I will take you to court. You have half an hour ... now, what did you want to know?

Gulp! Stay calm, Deepta, don't be derailed ... right, deep breath, here goes ...
And so I started the interview, which in the end lasted for two hours. With interruptions being waved away by the officer, he told me in minute detail all that I wanted to know and more. *Wow, he really knows this stuff ... and it is so interesting. But if I can't use this information, then what's the use of him telling me all this?* The interview was almost over when he said: 'You know, everything that I have told you is well documented. If you want to use that information, then you can get the documents and quote from them.'

Me: Sir, where are these documents?
Officer: They are all with me.
Me: Can I ask you to get access to these documents?

Officer: Well, you can have them but you will have to ask me officially for them, and see if you can get them.

With this, the interview was over. I drove home, mulling over what the officer had said. Given that my PhD would be incomplete and partially wrong if I omitted this material, I was naturally eager to get my hands on these 'official documents' which would help corroborate my story. But how was I going to do that? I wrote letters to the office of the Ministry where the documents were held, but to no avail. I pursued every lead so that I could triangulate the information I had, but no one in the government machinery would divulge anything. My ethical dilemma was that of having the most crucial information that I needed but not being able to use it without the officer's consent.

In order to resolve this dilemma, I started a long and arduous journey toward accessing these documents. Nothing had prepared me, however, for the intensity of the ethical dilemmas I would face during this journey, nor would I be able to reconcile easily the moral predicaments that the process of obtaining information would throw up.

I next met the same senior civil servant that I had interviewed at a conference a couple of months later. When I told him that I was having difficulty getting access to those documents, he smiled and said, 'it is your right to get the information, you might want to exercise that right.' I instantly realized that he was asking me to put in an application under the recently passed Right to Information (RTI) Act – and against his own department. This puzzled me: on the one hand the officer was saying I could not quote him, and on the other hand he was suggesting a legal route to obtain those documents. It seemed that although he wanted to share the documents with me, he was officially not allowed to, and hence he was suggesting a possible way around.

I was initially reluctant to put in an RTI application, as that would mean antagonizing other potential respondents with whom I was trying to seek inter-views and gain trust. However, pleading and pushing and trying other means had not succeeded, and I soon realized that the RTI was the only way to get access to this information. I tested the idea of such an application with both my activist friends and some government officials, most of whom reiterated that though this was the only route for accessing these documents, it could create a lot of antago-nism against me. After much deliberation and internal tussle, I submitted an initial application under the RTI Act to the Ministry concerned. I was confident that the RTI Act of 2005 which mandated government officials to provide almost all kinds of information (Government of India 2005) would make my task feas-ible. I did not imagine the intense resistance that I would face during the process, nor the extended time period that it would take to acquire the information.

I was very surprised when I had no reply to my initial application. After some time, I submitted a first appeal, as mandated by the RTI Act, to the Appellate Authority in the Ministry. This action heralded some information, but a large part of it was denied on grounds of it being vast and also confiden-tial. I was taken aback – this was the same office whose highest official had asked me to apply for the information. *What kinds of games was the official playing with me?*

It was only after a second appeal to the Central Information Commission (CIC) that I finally elicited a response from the Ministry regarding the remaining information. The Ministry invited me to inspect this 'exhaustive and voluminous' information and to make a selective list of the desired documents to be photocopied. During repeated visits to the Ministry over the next couple of months, I was asked several times to explain the reasons for my requesting this information. My response that it was for research purposes did not find favor with the junior officials who held these files and under whose scrutiny I had to peruse the documents. I encountered outright dismissal of research as a valid motive for operationalizing the RTI Act; in fact officials expressed disbelief that any Act would give a researcher the right to important information. Ultimately, as a researcher, I experienced personal hostility and was given the strong impression that I was wasting the time of hard-pressed government officials by asking for such information.

Despite these pressures I persisted with my application, but my request for photocopies of approximately 500 shortlisted pages was promptly rejected by the Ministry. This was on the grounds that they could not give me any information until the CIC gave their judgment pertaining to my second appeal. Meanwhile, I had also established informal contacts with a mid-level official in the Ministry, who advised me that I should submit a fresh application under the RTI Act as the previous one was under dispute and would not be easily resolved. In a show of support I was granted access to files to prepare an exhaustive list of documents which I could ask for in this fresh application, but was advised firmly not to ask for any cabinet papers and file notings. Taking their advice, and believing that this might result in my getting at least most of the information, I submitted a second application. However, given the importance of the cabinet notes for my work, I asked in this application also that these would be provided to me. This greatly annoyed the mid-ranking official, who turned quite hostile toward me. Again, my intent as a researcher to ask for these particular documents was questioned by both mid-ranking and junior officials, who repeatedly pointed out that these cabinet notes and file notings 'were of no use to anyone outside the ministry, especially a researcher.'

Simultaneously I had also written to the CIC stressing my time restrictions (my fieldwork timeline and impending return to the UK) and requesting an imminent hearing. Their response was positive: a date was announced for the hearing, at which myself and officials from the Ministry were called to present our respective cases. At this stage the mid-ranking officials asked me to strike a bargain: to take information pertaining to my second RTI application in exchange for withdrawing my appeal with the CIC regarding the first application. My refusal to accept only part of the information resulted in dirty politics by these officials, who were intent on pressurizing me to change my mind. I experienced personal intimidation through shouting, contradictory messages, and threats concerning the possibility of being denied access to even part of the information.

This scenario raised a dilemma between the objective and the ethics of my research. The objective of the research was to acquire maximum information within the constraints of time and resources. Here, time was limited, I had over-stretched my resources by repeated visits to the Ministry, and I was growing tired of fighting for this information. However, ethics demanded that I maintain the sanctity of the process, and above all that I not engage in 'deals' or other informal or unofficial methods to access the information. Was a compromise justifiable? For me, there was another important dynamic which related to an internal conflict around my positionality. Over the period of my fieldwork I had been interacting with various activists regarding this case, and I saw myself as ideologically and ethically aligned with them. I was therefore torn between the role of an impassive doctoral researcher and the activist who was not willing to make ethical compromises. It was at this crucial juncture that I realized the difficulty, yet the necessity, of combining academia and action. When I decided to place more importance on the process rather than the information itself, one mid-ranking official labelled me as 'being stubborn and spoiling her own case.'

This decision also alienated me from the mid-ranking and junior officials at the Ministry. An important moral predicament pertained not only to this case but also to my future relationships and access to information from the Ministry. By insisting on getting the information and taking the officials to court, was I not eroding any future relationships with the Ministry officials? *They will hate me for this and that is the end of any relationships with the Ministry for ever. I wanted to work with them, for them, and now I am doing myself a disfavor. Even if I win, I can never go to the Ministry again – at least not while this senior official is leading this division – that is the end of my work on the MGNREGA.* I was a bit apprehensive about the verdict too – if I lost the case, I would lose access to the majority of the information that was currently on offer too. *Then I would really be in trouble with my PhD.*

Assailed by self-doubts, I began reading books and articles on ethical positions. The work of Richa Nagar (1997, 2002) inspired me; the writings of Staheli and Lawson (1994) and Rose (1997) made me strong; and the writings of Dreze (2002) ensured that I persevered with my case rather than give in to the feelings of self-doubt. However, none of these works were set in a situation of researching with elite respondents. Therefore beyond all this literature, it was conversations with my researcher friends and supervisor that brought clarity and made my resolve stronger.

I also sought the help of an organization whose members were intimately connected to the activists involved in the establishment of the NREGA as well as the RTI. I went to the hearing at the CIC with one of the representatives of this organization. The atmosphere between the two parties was clearly hostile. The government officers – with many of whom I had spent hours talking, discussing, or poring over files – ignored me and pretended that I did not exist. The hearing at the CIC lasted no more than five minutes. The Central Information Commissioner rejected outright the Ministry's reasoning that the information asked for was voluminous, and that cabinet papers and file notings were confidential. The Commissioner upheld that the RTI Act (Government of India 2005) mandated that this information should be provided to me within 30 days. Since this commitment had not been met, the Ministry was directed to make available all information within the next two days and free of charge. The final blow to the officials came with an imposition of a fine for 8,000 rupees (250 rupees per day for 32 days, i.e., the number of days beyond the 30-day time limit during which the absent Public Information Officer had not given an answer to my application) that was to be imposed on the Public Information Officer directly.

The day after the hearing, I received all the information, without reservation, that I had requested. *I had all the evidence I needed!* I returned to the UK a day after obtaining the information and I heard that the CIC had held a showcase hearing with the Ministry officials, during which the imposed penalty charges were dropped because of the difficulty in pin-pointing the exact official on which to impose the penalty. *Well, at least the officials won't hate me for life, which they would have done if the fine had been imposed, as it was to come out of their salaries! But what about the senior official? He must surely hate me for having caused so much trouble and now humiliation for his department!*

Reflections on ethics and morals

Personally, the biggest gain for me in this process was the resolution of my internal conflict of identity between that of activist and that of academic; the academic's objectives could be fully achieved only through the activist's actions and perseverance. In this way, I have seen for myself how the disparate elements of research and action can be and, in fact, are bridged in real life, and how the seemingly distinct spheres of activist theorizing and academic theorizing are complementary rather than conflictual, as claimed by Baker and Cox (2001) (see also Jones in this volume on negotiating the dual role of researcher and practitioner in the field).

However, a few important ethical and moral questions remain to be discussed. One concern relates to the validity of research as a worthwhile activity. While collecting information, my intent as well as my legitimacy as a researcher was questioned at all levels – not only by the government officials, but also by friends and colleagues in social movements and other organizations. This has to be seen within the wider perspective of research being an inherently problematic activity, especially in India where research and academics are generally viewed critically, if not suspiciously. The formal order of the hearing vindicated my doctoral research as a legitimate and important activity:

'... the conduct of a study of the decision making processes ... of a flagship program such as the NREGA is to be appreciated in a proper perspective. It is unfortunate that the effort made by a researcher to undertake an in-depth analysis has been discouraged by withholding information on the pretext of voluminous information'.

(legal judgment made to the author)

This can be seen as a historic and trend-setting judgment; this was the first time that the CIC had upheld the importance of research and the use of the RTI Act (Government of India 2005) for this purpose. However, this cannot be taken as a validation for all research and every researcher, and certain caveats remain.

First of all, the RTI Act (Government of India 2005) recognizes the right to information for any Indian citizen. It does not recognize the same right for foreign researchers to access information through the use of the Act. This has implications for research which requires information that can only be accessed through the RTI Act. Does this mean that such research can be undertaken only by Indian citizens or by researchers who have Indian collaborators?

Secondly, the use of the RTI Act as a tool for research needs to be judicious and justifiable. While there is a potential to use the Act to get access to information for purposes of research, I should add a note of caution about its future use and the prevention of its overuse. The intent of the Act is to make governments more accountable to the people that they serve. However, already overburdened public officials may not have a way of sorting through a plethora of RTI applications to ascertain which ones are relevant or are a judicious use of their time and resources. It then becomes the responsibility of the researcher to use the RTI Act judiciously, and to weigh the benefits of their purpose as compared to other purposes such as social audits of muster rolls and wage payments before deciding whether to use the RTI as a research tool. This is one of the key predicaments that I needed to work through. *I still wonder whether my use of the RTI to get information was justified. How many other applications were put aside while mine took time and energy of the government officials to sort out? How many other files remained unseen and unprocessed while I harassed the officials for information that I needed for my research?*

I resolved this dilemma by promising to myself that I would be diligent in ensuring that the information collected was put to good use. This then landed me in another set of ethical questions about the purpose of generating knowledge – this can be either on the basis of its practical effectiveness in bringing about change or for generating contemplative theoretical knowledge. The RTI Act (Government of India 2005) recognizes that information provision can 'conflict with other public interests, including efficient operations of the Governments, [and] optimum use of limited fiscal resources.' Given that asking for information takes up public resources, as a researcher I then have had the responsibility to make this information public or employ this information for a public purpose. *If my research serves only a private purpose, i.e., the completion of my doctoral thesis, exercise of the RTI Act is not justified.* Only when research as an activity serves a public purpose can the use of RTI Act be validated.

Ultimately, it was my responsibility to ensure that the findings of this research were made public and accessible and that the knowledge generated serves a practical purpose. I offered to provide photocopies of the documents I obtained to activists and other researchers working on this subject. Above all, I ensured complete anonymity of all respondents in all my writings. This was not an easy undertaking given the high profile and easily-identifiable nature of many of the respondents and the politically-charged nature of my topic. However, I remain reticent in sharing these documents publicly, and am aware of their sensitive and political nature. In fact I denied access to a newspaper reporter just after receiving these documents because I was not convinced of their intent in using these documents. *Have I then become a gatekeeper myself? Is it up to me to decide who gets access and who does not get access to these documents? Where does my responsibility lie – in making sure that the documents are not misused, or in protecting my own evidence base?*

Another partially resolved moral predicament has been my relationship with Ministry officials. I have been observing the happenings of the Ministry closely but without setting foot near the place. Despite the fact that the senior civil servant pointed me in the right direction for obtaining the information, I have not yet been able to forget either his intimidating nature nor the hostility of his department officials. I heaved a big sigh of relief when the official was transferred. My fear of being hated and remembered as the trouble-maker for the department was only assuaged by a later trip to the Ministry, where I found lots of new faces and no sign of any hostility even among the couple of familiar faces that I saw.

Becoming an 'academivist'

In this chapter, I have reflected on a year-long journey to access information from the Indian Ministry of Rural Development through the use of the Right to Information Act. Through this story, I have illustrated how a conflict between being an academic and being an activist came alive for me. To do this, I began with my initial position as a doctoral researcher, i.e., a prudently distant individual concerned with using the RTI Act as a research tool. I then related how this positionality of an academic self was compelled to move toward an action-oriented participant who is both studying and being studied as part of the process of data collection. Through reflecting on this experience, I have examined what the process of becoming an 'academivist'[2] entails (Morell 2005) in terms of representing myself as someone who is part of an academic institution, but because of some personal and/or professional strategy is also part of a social movement.

My experience suggests that it is not only possible but in fact essential to combine the roles of academic and activist into what can be termed as an 'academivist' positionality. I agree with Dreze when he says that

> [t]he value of scientific research can, in many circumstances, be enhanced if it is combined with real-world involvement and action. This approach

should be seen as an essential complement of, not a substitute for, research of a more 'detached' kind.

(Dreze 2002)

In addition, I would highlight that in order to be legitimate and useful to both academics and activists, this process of combining the worlds of research and action requires research (and academicians) to be responsible, justifiable for a public purpose, and, above all, dealing with ethical dilemmas and moral predicaments. The process of becoming an 'academivist' is then not simply combining the identity and working style of the academic and the activist. It also necessitates the application of knowledge in the generation of practical change, and in dealing with ethical and moral issues in the most honest way that you can.

Recommended reading

Chacko, E. (2004) 'Positionality and praxis: Fieldwork experiences in rural India,' *Singapore Journal of Tropical Geography*, 25 (1): 51–63.

This is an insightful article based on Chacko's fieldwork experiences wherein she explores the impact that power has on both the research process and the relationships between the researcher and the researched. Her candid and reflexive writing about situations that are 'simultaneously familiar and foreign' is very helpful. The article talks about 'building and maintaining relationships' in the field in order for research to not be purely extractive.

Dreze, J. (2002) 'On research and action,' *Economic and Political Weekly*, 2 March, 37 (9).

This article reflects on the challenges associated with doing research in situations where the researcher is also an activist or is closely associated with social movements and organizations which may be engaged directly in specific situations, something that Dreze calls 'real world involvement and action.' Dreze highlights not only how research contributes to this real world involvement, but how involvement in action enhances the quality of the research.

England, K. V. L. (1994) 'Getting personal: Reflexivity, positionality and feminist research', *Professional Geographer*, 46 (1): 80–89.

This article speaks about power relations at play in field work, highlighting some of the key ethical challenges facing researchers, with a focus on the relations between the researcher and the 'researched' – or those who are interviewed. This is a very useful article in highlighting the role that the researcher's (and the interviewee's) positionality plays in structuring and shaping the research process.

Richards, D. (1996) 'Elite interviewing: Approaches and pitfalls,' *Politics*, 16 (3): 199–204.

This is a very clear and succinct introduction to the technique and challenges of elite interviewing, which is quite different to interviewing non-elites. I particularly liked the way in which Richards talks about the need for flexibility – a lesson that came in very handy while negotiating the interview terrain with highly placed and senior civil servants, but also with other government officials.

Notes

1 'Peon' refers to a male usually charged with carrying files, guarding the entrance to the office, answering to calls for tea and coffee, ushering in visitors, and running small errands in and outside the office.
2 The term 'academivist' has been used by Morell (2005). She defines academivists and 'academivism' as 'an ideological corpus and individuals that being active at social movements and from a critical approach to the academic institution, but researching from its framework, search to find ways to contribute to the movements with the research "about" movements they are developing.' I am using the term 'academivist' conversely, i.e., individuals having a primary institutional affiliation with academia, but taking on roles of a social movement activist during the research process.

References

Baker, C. and Cox, L. (2001) '"What have the Romans ever done for us?" Academic and activist forms of movement theorizing,' available at: www.iol.ie/~mazzoldi/toolsfor-change/afpp/afpp8.html (accessed 8 June 2008).
Bernard, H. R. (2002) *Research methods in anthropology: Qualitative and quantitative approaches*, Thousand Oaks, CA: Altamira Press.
Dreze, J. (2002) 'On research and action,' *Economic and Political Weekly*, 2 March, 37 (9).
Elwood, S. A. and Martin, D. G. (2000) '"Placing" interviews: Location and scales of power in qualitative research,' *Professional Geographer*, 52 (4): 649–657.
Government of India (2005) *Right to Information Act*, Delhi: Ministry of Law and Justice, Government of India.
Herod, A. (1999) 'Reflections on interviewing foreign elites: Praxis, positionality, validity, and the cult of the insider,' *Geoforum*, 30 (4): 313–327.
Hert, R. and Imber, J. B. (1995) *Studying elites using qualitative methods*, Thousand Oaks, CA: Sage.
McDowell, L. (1998) 'Elites in the city of London: Some methodological considerations,' *Environment and Planning A*, 30: 2133–2146.
Morell, M. F. (2005) 'Political action and investigation interactions today: Invitation to action research the Social Forum Process,' available at: www.euromovements.info/html/mayo-foros-sociales.htm (accessed 8 June 2008).
Nader, L. (1974) 'Up the anthropologist: Perspectives gained from studying up,' in D. H. Hymes (ed.) *Reinventing Anthropology*, New York: Vintage Books.
Nagar, R. (1997) 'Exploring methodological borderlands through oral narratives,' in J. P. Jones III, H. J. Nast and S. M. Roberts (eds), *Thresholds in feminist geography: Difference, methodology, representation*, Maryland: Rowman and Littlefield Publishing Inc.
Nagar, R. (2002) 'Footloose researchers, "travelling" theories, and the politics of transnational feminist praxis,' *Gender, Place and Culture*, 9 (2): 179–186.

Razavi, S. (1993) 'Fieldwork in a familiar setting: The role of politics at the national, community and household levels,' in S. Devereux and J. Hoddinott (eds), *Fieldwork in developing countries*, Boulder, CO: Lynne Reinner Publishers.

Richards, D. (1996) 'Elite interviewing: Approaches and pitfalls,' *Politics*, 16 (3): 199–204.

Rose, G. (1997) 'Situating knowledges: Positionality, reflexivities and other tactics,' *Progress in Human Geography*, 21 (3): 305–320.

Staeheli, L. A. and Lawson, V. A. (1994) 'A discussion of "Women in the Field": The politics of feminist fieldwork,' *Political Geography*, 46 (1): 96–102.

20 Restaurants and *renqing*: ethical challenges of interviewing business people over dinner

Experiences from fieldwork on TNC supply networks in China

Yue Wang

It was only May but Shanghai was already very hot. I was sweating as I stood outside the shopping mall waiting for the food regional manager. 20 minutes gone. He just told me to meet him here but I had never met him before and didn't know what he looks like. His mobile phone was turned off. Would he show up? Should I wait here for another 10 minutes? Oh, a phone call! 'Ah, your mobile phone was out of power and you just got it charged. Yes, I will come to the restaurant in the shopping mall immediately. It's on the ground floor? OK, I will find it. Thank you. Thank you so much!'

When interviewing professional people you would expect a scheduled meeting booked into a calendar, and the researcher and interviewee to be seated at a table with the essential equipment of dictaphone, notebook, and pen. You might also have tea and cookies as a mark of hospitality. In my case, add to this situation a revolving table, dishes of food, chopsticks, and a hovering waiter, because a significant number of my interviews with business people in China took place in restaurants over dinner (as arranged in the account above). This chapter explores some of the ethical and practical challenges raised by this scenario.

My doctoral research examined how the arrival of retail transnational corporations (TNCs) in China has transformed the supply network and upgraded the market. My research was carried out in Shanghai, my native city. During my first phase of fieldwork I contacted retailer representatives and food suppliers in order to gain background information on the Chinese retailing market, as well as to identify the appropriate food stuffs for my case studies. During the second phase of fieldwork I identified specific procurement shifts adopted by retail TNCs and in turn the responses made by suppliers/wholesalers and logistics providers. This focused on three selected food types: fresh apples, fresh milk, and own-branded edible oil. In total I carried out 54 semi-structured interviews with a range of people including leading retail representatives, food suppliers/ wholesalers, logistics providers, academic scholars, business consultants, and governmental officers.

The most important types of informant – yet equally the most difficult to access – were retailer representatives, logistics providers, and food suppliers. By the end of my fieldwork I felt that I had successfully achieved my aims in terms of talking with the right kind of people about my research topic, but this process was neither simple nor smooth. Dealing with business people raised a number of issues about power relations; add to this the fact that many interviews took place in the setting of a restaurant where there was often an ambiguity about whether I, the researcher and interviewer, or the interviewee was actually in charge or acting as host. Being in a restaurant setting posed a number of ethical challenges regarding the choice of venue and choice of food, behavior while simultaneously eating and conducting an interview, and the issues of paying the bill and farewell requests. This chapter will explore each of these aspects of conducting interviews with business people in restaurants. The examples from my experience should be of relevance to other researchers doing fieldwork in China which involves elite interviews, as well as researchers in other countries where the interview setting may involve issues of hospitality and hosting.

Yaoqing (the invitation)

I started my quest to find relevant business people with an intensive internet search. Having identified potential participants I then contacted them via email or telephone. Once I had established some key contacts I then planned to use snowballing techniques to identify other potential informants. However, the cold-calling approach was largely unproductive and frustrating. In most cases I had either no response or direct refusals via telephone. The likely reason is that people in the retail sector are very busy; if they agreed to take part in my research project it would take their time and I could offer little in return to compensate them.

Having largely failed with cold-calling I turned to my personal social networks. Fortunately I have lived in Shanghai for over 20 years and therefore have established relatively strong social networks locally. Thus I contacted my family and friends, previous colleagues, and university alumni in an attempt to build up my interviewee list. In many cases my contacts passed on the request to their own social networks; for example, one of the important retail representatives was accessed via the husband of one of my alumni contacts.

Using social networks meant that the process of searching for potential interviewees was somewhat out of my control and certainly not random. However I still needed to strive for some representativeness across my interviewees. I did give my personal contacts very clear instructions in terms of the kind of attributes that I was looking for in my interviewees. I also requested them to refer informants from different companies wherever possible, thereby avoiding recruiting informants centered around certain organizations (Browne 2005) and enabling me to talk with a wide range of people with differing experience and specialized knowledge. For example, with the food product of fresh milk my personal networks allowed me to reach people from the top three dairy companies. This enabled me to look across the milk sector and not focus on a single company's practices.

In total, almost half of my interviewees were reached on the recommendation of my friends and family members. Using personal social networks to identify and contact potential interviewees or participants will raise some ethical issues for any researcher (see also Taylor in this volume on using family members). In my case and in the Chinese situation, using personal social networks created a very particular ethical dilemma surrounding reciprocity which I will describe and explore later in this chapter.

Canting (the restaurant)

Whether my informants had agreed to be interviewed via social network referrals, snowballing by other interviewees or cold-calling, I then needed to arrange a time and place for the interview. Knowing that these were busy professional people, and that I was very fortunate to have been granted an interview, I decided to be very flexible in allowing them to choose the interview time and location. Almost one-third of my respondents chose to meet me over dinner or snacks after their working hours. I assume that this decision was due to three reasons. First, people working in the retail sector are very busy and would not want to spare their valuable working time or lunch breaks to talk with me about matters that were not directly related to their everyday business. Second, they may have been concerned that if they had been interviewed during work time then colleagues might have thought that they were not working (McDowell 1998). Third, and perhaps most importantly, giving a feast (*qingke chifan*) is one of the most effective ways to build rapport with people in Chinese society (Hwang 1987; Zhuang *et al.* 2008). By suggesting meeting in a restaurant, respondents may have been trying to show their hospitality and respect to me. On the other hand, they may have expected me to show my hospitality and respect to them by paying the bill, which is an issue I will come to later.

The negotiations about the location of the interview often took some time. Although I always offered my interviewees freedom of choice, some of them asked for my preferred meeting place. This might have been driven by the concern that if the proposed place was too far away from my home I would have to spend too much time travelling. Alternatively they may not have been sure whether their suggested restaurants would be too upmarket or too casual for my research interviews. Whenever I was provided with the opportunity to choose the interview venue, I suggested a medium-sized restaurant with affordable prices but professional service. Here I attempted to achieve a balance between the interview quality and related costs. On the one hand, this kind of restaurant would not let my interviewees feel that my research was casual or informal, and in turn lead to unsatisfactory interview quality. On the other hand, I could afford the bill to show my hospitality.

In this whole process was a bargaining-power shift between my informants and myself. That is to say, my interviewees had relatively more power to decide whether to take part in my research as well as when and where the interview would take place. However, the power was not always in their hands; rather it was diffuse and mobile. This notion is in line with Allen's (2003: 8) conception of power as 'only ever mediated as a relational effect of social interaction.' It is widely recognized that elite people are more difficult to get access to than other groups (Cochrane 1998; McDowell 1998; Parry 1998; Sabot 1999) and in turn they may have more influence on how the interview should be conducted. However, from my personal experience, researchers can still bargain with interviewees to reach a compromise in terms of interview arrangements, such as meeting time and venue.

Canzhuo (the table)

Although private rooms are quite common in China I did not choose to conduct my interviews in this setting. Private rooms usually require minimum dining charges which would have been out of proportion to a meal involving just two guests, so we sat in the main part of the restaurant. When I greeted the interviewee at the selected restaurant, I gave them the choice of table but wherever possible I advised that we should choose one in a quiet location. There were several reasons for this preference. First, in practical terms, it would enable a higher-quality recording of the conversation away from the noise of other customers and reduce any potential distractions or interruptions. Second, in terms of confidentiality, I was concerned that my interviewees might not speak with me freely and frankly if they were afraid of being overheard by people at neighboring tables. However, to my surprise, few of the interviewees seemed worried about this issue and most claimed that what they had told me was simply the daily routine of their work, which was not a secret at all within the industry. In fact, all except one informant agreed to the interview being recorded. As a result of this I realized that my research topic actually posed little concern in terms of sensitivity. This also enhanced my confidence in asking some difficult questions in a relatively relaxed atmosphere. Overall, neither the presence of other customers situated close to our table nor that of the waiter/waitress who served us made much impact on the interview process.

There was one awkward situation, though, involving a businessman who brought his girlfriend along to the restaurant. It was the only interview which took place with a spectator. As his girlfriend had little interest in my research project, she occasionally showed her impatience with the interview process, which inevitably distracted both the interviewee and me from our conversation from time to time. Finally I had to propose stopping the interview earlier than I had planned, as the lady insisted on going home. Although the interviewee gave me some useful material, I was still perplexed about his girlfriend's behavior. I wondered whether the interviewee had utilized his girlfriend as a barrier to my interview because he had little interest in my research but was embarrassed to refuse his friend's request to attend the interview. Maybe this was a very particular example, but in the following interviews I tried my best to ensure that only the informant and I were present in order to avoid any influences from third-party attendees.

Caidan (the menu)

By asking an interviewee's preference in picking a particular table, I was showing my respect to them and starting to build a rapport with them. This process was in turn strengthened when it came to looking at the menu, which involved further 'warm-up' conversation before the formal questioning of the interview started.

In Chinese society it is the traditional convention for the host and guest to use communal bowls. However, this practice of sharing food usually takes place among families and friends and shows a close sense of intimacy. In our case we selected our own choice of dishes from the menu, as we were not so 'close' to each other. This maintained a proper distance between us and avoided any potential embarrassments of taking the same food from one dish. It also avoided any issues that might have arisen if one of us had particular dietary preferences. Researchers in other cultures may find that sharing from the same plate is actually crucially important to building a bond and developing trust, but in my case ordering separately was an important way to be respectful and maintain a professional distance.

Drink can also potentially present an ethical dilemma to the researcher conducting interviews in the setting of a restaurant, café, or bar. In Chinese society, drinking culture plays a very important role in building business relationships (Hao *et al.* 2005). However, it is also a tradition for young females to take little drink in a public setting (Newman 2002). Furthermore, I am not accustomed to drinking alcohol, particularly with strangers. Although social drinking might have been useful to build rapport with my business informants, I did not think that my research interview necessarily fitted with this cultural norm as my interviews were not official business meetings which required considerable social drinking to facilitate a negotiation or long-term collaboration. On the contrary, my interviews were relatively short-term collaborations between myself and the informant and we had few mutual interests in each other.

Having judged that social drinking was not an essential component of my interviews, even with business people, I decided that if I was acting as host, I would not ask my interviewees whether they wanted alcohol but instead suggest ordering soft drinks. In fact, my interviewees did not ask to drink alcohol with me. This was probably because it would have been regarded as a bit impolite to do this with a lady that they were not familiar with. As with food, researchers working in other cultures must judge what the most appropriate behavior is regarding taking alcohol and assess the ethical issues that rest with it.

Yongcan (the meal)

When the food was served, my native knowledge of Chinese table etiquette was very important in showing respect. If I was acting as host I asked the interviewees to sample the food first in order to show my hospitality to them; if they were acting as host I waited to take food after them, also to show my respect. In China the host also asks the guests to sample their food in order to show their consideration for the guest. I recommend that a researcher who is doing fieldwork outside their own culture takes time to learn some of the basic behavior and etiquette of that place in order to avoid embarrassment or causing offence at the table.

Once eating, I felt that people tended to talk more freely with me during the meal and this also gave me more confidence to ask difficult questions and request other contact names. However, I want to warn that combining interviewing and dining is a difficult task (Dexter 2006; Harvey 2009). I was concerned that people might not concentrate on my questions when they were taking food and that this would affect the quality of my interviews. There was a tricky balance between enjoying the food and focusing on the questions. Sometimes I had to remind my interviewees to take food if they kept talking for a long time but ate little. I also had to push them to clarify their arguments with more detail on some occasions. Meanwhile I did not allow myself to fully indulge in food so that I could focus on controlling the conversation and keeping notes of what I was particularly interested in. These strategies helped to facilitate the interview process to be more formal and also to minimize the distractions from food to my interviewees as well as to myself.

As mentioned earlier, one practical concern was that a noisy restaurant setting would affect the quality of the interview recording. In fact, I found that this concern was minimized by placing the recording machine closer to the interviewee than myself and away from moveable objects such as drinking cups and chopsticks. Also the machine had a very good recording quality, and once I played back the first few interviews I realized that I did not need to worry about conducting interviews in a noisy environment.

Although background noise was not a distraction, it was not unusual for my informants to receive and answer phone calls during the interview (see also Chopra in this volume). For example, during a 40-minute interview with a Managing Director of food suppliers, she received no less than three phone calls. Each time I had to stop my recorder to avoid recording irrelevant material (to me) or confidential content (for her) and make notes of where I needed to resume later. It was really a frustrating process when it kept happening, and each portion of my interview lasted for only five to ten minutes. The interview did not flow as I expected and also my confidence to carry on with the interview was affected. On the one hand, I knew that I needed to ask a question immediately after the phone call to get back to my topic. On the other hand, the phone call interruptions made me feel under pressure that I was occupying her precious time and was probably a troublesome annoyance in her schedule. Other researchers are likely to face this problem, particularly when interviewing busy business people or other elites, no matter what the interview setting is.

To be honest, I would have liked to ask my interviewee to switch her phone off or ignore calls, but I was pretty sure that making this request would be considered impolite. The interviewee was effectively my guest, as I had invited her to talk with me. As a result, it would be inappropriate to require the guest to do something only favorable to myself but against her will. After all, the participation in my research was voluntary and the informant had the right to withdraw at any time she wanted. I therefore did not want to make her feel annoyed by any of my requests and in turn affect the interview progress. Thus, in my case I decided to persist with my questioning in order to obtain as much information as possible in the time available.

Maidan (the bill)

It is common for any mixed group sharing a meal, whether in a social or professional setting, to face a moment of discomfort regarding who is paying the bill. There may be assumptions by one party or the other about who is acting as host. As I explained earlier, in my interviews with business people it was sometimes ambiguous whether I or the informant was actually the host in the restaurant setting. When it came to paying the bill, I would like to highlight the very important role played by *renqing* (human feeling) in the Chinese setting.

Renqing refers to a set of social forms that the Chinese value highly and which keep good relationships with other people (Wang *et al.* 2008). According to Hwang's (1987) conceptual framework of social behavior in a Chinese society, people make decisions in terms of their various relationships or ties (*guanxi*) with others, which may be instrumental ties or mixed ties (see Figure 20.1).

As I explained earlier, many of my informants had been identified via social networks. People connected to one another in these kinds of mixed ties tend to perform their social conduct based on the rule of *renqing*. Both sides of the mixed tie know each other and have something in common with one or two persons; so both would like to make contributions to each other in order to maintain *renqing*. Thus, in my fieldwork experience, my social networks were willing to help me look for informants as they thought they had obligations to keep *renqing* with me. Similarly, those informants agreed to share their knowledge

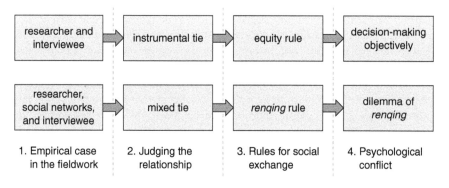

Figure 20.1 Relationships between the researcher and interviewees in Chinese society (source: Adapted from Hwang 1987, Figure 1).

and time with me for the sake of their *renqing* to the people who contacted them, rather than out of any sense of obligation to me.

Before each interview I was prepared to pay for the meal, not only as I had requested people to spend time talking to me but also because I needed to repay *renqing* to the interviewees on behalf of my social networks. Paying the meal bills was effectively equivalent to repaying *renqing*. However, I was only a post-graduate student and had not been able to gain any additional funding for my fieldwork, so I had to budget for the costs of the restaurant bills from within my doctoral maintenance grant. That was also a reason why I proposed reasonably-priced restaurants whenever the informant wanted me to choose the interview venue.

However, I did not always pay the restaurant bill. Some interviewees paid for the meal even if I had expressed my willingness to pay. Their reasons for this were varied. For example, some of them mentioned that they had decent jobs and would not allow a student without income to pay the bill for them. In other cases I had been recommended by very good friends of theirs and they would lose face to their friends if I paid the bill. In this case *renqing* again was highly significant because the relationship between the referrer and the interviewee was critical to the whole interview experience. Not surprisingly, if the referrer was in a relatively higher position or had close relations with the interviewee, I could sense more hospitality and cooperation during the interview as well as regarding the bill. For example, I once talked to a Regional Food Director of a leading retailer for around an hour without any interruption from his phone calls. He also gave me in-depth answers to my questions and was glad to elaborate further when I inquired. In fact, he had been referred by his supervisor, the Expansion Director in headquarters, although I had no idea about the Expansion Director at all.

Gaobie (the farewell)

In my case the unspoken yet highly significant socio-cultural framework of *renqing* not only provided me with a way to negotiate the ethical dilemmas of who should pay the restaurant bill, but also remained important as I wrapped up the interview and prepared to say goodbye. As mentioned earlier, I was trying to use the snowballing technique to reach more informants, so I wanted to ask interviewees to recommend people from among their contacts. However, my interviewees were free to make their own decision whether to give me an extra 'gift' (i.e., further contacts), as they did not owe me an obligation of *renqing*. For example, one of the interviewees declined to give me further contacts by stating that he attended the interview only for the sake of his friend (i.e., the referrer). Although I felt disappointed with refusals of my requests, I fully under-stood the reasons behind them.

However, in cases where I paid for the meal the situation changed subtly in favor of me. One of the basic rules of conducting *renqing* in Chinese business circles is reciprocity. If someone receives a favor, he or she owes *renqing* to the benefactor, and should be ready to pay back once circumstances permit (Wang 2007). By paying the bill I had not only repaid *renqing* on behalf of my personal networks but also gained the opportunity to ask for any further contacts. Some interviewees might have felt that they then had the social obligation of reciprocity to me as I had paid for the meal; thus they were embarrassed to say no immediately and usually replied that they would think about it later.

After the interview, I sent a thank-you email to my interviewees and again appealed for further potential informants from among their contacts. If I did not receive any response within two weeks I also made a phone call. I did succeed in getting some positive responses using this method and was given further contacts. Here I do not think that these people helped me because I had paid for the meal, but this fact had given me more confidence to push them for any further contacts because I tried to make them feel that they somehow owed me *renqing*. Nonetheless, I was not trying to deliberately manipulate *renqing* over my interviewees for my research purposes. Once my interviewees declined to offer further help when I contacted them again, I showed my understanding of their decisions (although sometimes not very rational explanations) and my appreciation of their involvements.

I would like to give one more example of *renqing*'s complex impacts on my fieldwork. Eight months after my fieldwork I received an email from one of my informants who wanted to refer to my research results for his Master of Business Administration thesis. This interviewee was one of the few who had paid for dinner, and he had given me another two contacts for my subsequent interviews. In my heart, I really appreciated his help and also felt that I still owed *renqing* to him as I had not found a way to repay his help. At that time, I was still in the writing stages of my research so I sent him my first-year report for his reference and promised to give him the whole draft of my thesis when it was completed. Here *renqing* had helped me build long-term relationships with my interviewees for the foreseeable future.

To sum up, *renqing* is an extremely important and dynamic factor to be taken into consideration in the Chinese setting. As a native Chinese person I fully understood its role and nuances, particularly in business relationships. While I used it to gain access to interviewees, it also sometimes made me embarrassed to push them further with difficult requests as they did not owe me anything else. Since *renqing* involves reciprocal relationships, it may mean both pros (e.g., more contacts) and cons (e.g., monetary costs) for the researcher depending on the situation.

Who is playing host?

In this chapter, I have reviewed my experiences of interviewing business people in China, specifically in the setting of a restaurant. I reflected on the different stages of the process including the initial contact and invitation for an interview, the choice of restaurant and choice of table, decisions over the menu about food and drink, behavior while eating and simultaneously interviewing, and then issues surrounding paying the bill and asking for further favors. In the relationship between myself as the researcher and the business person as the interviewee were negotiable power relations and a range of ethical dilemmas, as other researchers studying elite or professional groups may find (see also the chapters by Chopra and Perera-Mubarak in this volume).

There was an ambiguity about who was actually acting as the host in the restaurant setting. I allowed them to choose things (e.g., the time, the restaurant and the table) while still trying to control aspects of the situation. There was always a careful path to tread between formality and respect at the same time as hospitality and openness. Furthermore, the dynamic role played by *renqing* significantly influenced the behavior of interviewees in accepting interviews, in the openness of their responses, and in their willingness to help further by providing additional contacts. As a Chinese native, I found that my understanding of sociocultural norms such as *renqing* and Chinese table etiquette was invaluable in dealing with ethical challenges appropriately and sensitively.

As I noted above, conducting interviews in a restaurant over dinner is not easy, either in a practical sense or in terms of the ethical challenges. However, overall I found it an effective way to collect useful information from business people. Meeting at a restaurant and spending time discussing the choice of table and the menu was a time to build rapport. Then the dining setting provided a relaxed environment for interviewees, which enabled detailed explanations and clarifications when I probed for them. I also think that my interviews spent longer in a restaurant than they would have in an office. For example, I interviewed a logistics provider for around two hours over dinner, which would have been highly unlikely had the interview been held at his work place. However, conducting interviews over dinner was not without hazards or problems. Practical considerations included recording an interview in noisy restaurant surroundings and asking questions and taking notes at the same time as eating and drinking. Ethical issues included the confidentiality of conversations in a public place.

Sharing a feast plays an important role in building relationships in Chinese society. Researchers working in other countries and cultures may face similar situations where they need to offer or receive hospitality involving food and drink. I recommend thinking about the practical and ethical issues that I have described, and also being familiar with the particular cultural practices and etiquette of their particular fieldwork location to avoid embarrassment over issues such as eating from the same dish or drinking alcohol. Finally some common sense is needed, for example being sensitive to the balance between questioning and eating.

Recommended reading

Chen, V. (1990) '*Mien tze* at the Chinese dinner table: A study of the interactional accomplishment of face,' *Research on Language and Social Interaction*, 24: 109–140.

Gives a detailed analysis of Chinese table etiquette and also examines the host-guest relationship in dining settings; particularly useful to those who are not familiar with Chinese culture.

Harvey, W. S. (2010) 'Methodological approaches for interviewing elites,' *Geography Compass*, 4: 193–205.

Provides useful practical guidance on preparing for interviews and gaining access; also discusses the associated power relationships.

Lai, K. (2007) *Approaches to 'markets': The development of Shanghai as an international financial centre*, unpublished thesis, China Policy Institute, University of Nottingham.

The methodology chapter reflects on the researcher's experience, as an ethnically Chinese person from Singapore, of conducting interviews with both foreign and local business people in the financial sector in Shanghai, China.

McDowell, L. (1998) 'Elites in the City of London: Some methodological considerations,' *Environment and Planning A*, 30: 2133–2146.

Discusses a number of methodological issues in terms of interviewing elite people including getting access, interview location, interview processes, and final writing up; written from a feminist geography perspective.

References

Allen, J. (2003) *Lost geographies of power*, Oxford: Blackwell.

Browne, K. (2005) 'Snowball sampling: Using social networks to research non-heterosexual women,' *International Journal of Social Research Methodology*, 8: 47–60.

Cochrane, A. (1998) 'Illusions of power: Interviewing local elites,' *Environment and Planning A*, 30: 2121–2132.

Dexter, L. A. (2006) *Elite and specialized interviewing*, Colchester, UK: ECPR Press.

Hao, W., Chen, H. and Su, Z. (2005) 'China: Alcohol today,' *Addiction*, 100, 737–741.

Harvey, W. S. (2009) 'Methodological approaches for junior researchers interviewing elites: A multidisciplinary perspective,' Economic Geography Research Group Working Paper Series No. 01.09: 1–33.

Hwang, K.-K. (1987) 'Face and favor: The Chinese power game,' *American Journal of Sociology*, 92: 944–974.

McDowell, L. (1998) 'Elites in the City of London: Some methodological considerations,' *Environment and Planning A*, 30: 2133–2146.

Newman, I. (2002) 'Cultural aspects of drinking patterns and alcohol controls in China,' in D. Rutherford (ed.) *The Globe*, London: Global Alcohol Policy Alliance.

Parry, B. (1998) 'Hunting the gene-hunters: The role of hybrid networks, status, and chance in conceptualising and accessing "corporate elites",' *Environment and Planning A*, 30: 2147–2162.

Sabot, E. C. (1999) 'Dr Jekyll, Mr H(i)de: The contrasting face of elites at interview,' *Geoforum*, 30: 329–335.

Wang, C. L. (2007) 'Guanxi vs. relationship marketing: Exploring underlying differences,' *Industrial Marketing Management*, 36: 81–86.

Wang, C. L., Siu, N. Y. M. and Barnes, B. R. (2008) 'The significance of trust and renqing in the long-term orientation of Chinese business-to-business relationships,' *Industrial Marketing Management*, 37: 819–824.

Zhuang, G., Xi, Y. and El-Ansary, A. (2008) 'The impact of interpersonal guanxi on exercise of power in a Chinese marketing channel,' *Journal of Marketing Channels*, 15: 185–210.

21 Can you please all of the people some of the time? Ethical challenges in making research relevant to academia, policy and practice

Experiences from working with an international NGO in Mali

Stephen Jones

'You're a de facto member of the WaterAid team now.'

This was how the organizer of a WaterAid workshop described my status among the staff of WaterAid Mali about halfway through my fieldwork. In many ways it captures some of the ethical challenges that I faced in balancing the demands of academic fieldwork with my commitments to the partner organization involved in my doctoral research. My research was funded by the Economic and Social Research Council as a Collaborative Award in Science and Engineering[1] (CASE) studentship, with Royal Holloway, University of London as the host institution and the international NGO WaterAid as the non-academic partner. My own background before starting the studentship was not in the social sciences, but rather in engineering and NGO management, a position I discuss later in this chapter.

The research project's original title was *Improving local-level governance to meet the Millennium Development Goals for Water and Sanitation: The case of WaterAid in Mali*. This initial broad proposal was developed by the project's supervisory team, comprised of academics from the geography department at Royal Holloway and representatives from WaterAid in the UK and Mali. The project was designed as an opportunity to work with WaterAid in examining the organization's approach to working with communities and local governments in Mali, where decentralization reforms have left local governments with the responsibility for ensuring drinking water services, but a lack of experience and resources with which to do so. The intention was that the scope of the research could then be narrowed down further during the course of the doctorate. Much of this process took place during an 18-month period which I spent based in the WaterAid office in Mali.

The research topic that emerged was an analysis of how the recurrent costs of rural water supplies (operation and maintenance, technical and management support, and eventual rehabilitation of old systems) are shared between different actors including the users, WaterAid and its partner organizations (local NGOs and local governments), and national government. Financing arrangements for recurrent costs are a crucial element in ensuring the long-term provision of rural water services, but few countries have yet developed adequate frameworks to plan for and allocate these costs (Lockwood and Smits 2011).

This chapter explores some of the ethical challenges that I faced before, during, and after my fieldwork period as an academic researcher embedded within an NGO. I discuss the careful negotiation of the exact research topic that was necessary to ensure that it would benefit my own research needs as well as the organization and the communities it worked with. I explore the difficulties of having an ambiguous position which was simultaneously researcher, consultant, and practitioner. I also explore the complexities of seeking to ensure that my research would be relevant to academia, policy and practice.

The ethical challenge: combining research for knowledge, policy, and practice

The main ethical challenge I encountered in doing research with WaterAid in Mali was how to balance the need for the research to respond to three potentially competing demands: 'generating knowledge, informing policy or guiding practice' (Cleaver and Franks 2008: 165). Cleaver and Franks propose these three categories based on their own research in the water sector to help explore the possible tensions within the range of academic activities which are termed 'research,' and to suggest how these might differ in terms of scope, focus, timescale, type of data and presentation of results, and audience. Although acknowledging that the boundaries between research for knowledge, research for policy, and research for practice can sometimes be unclear, I still think it is useful to consider which parts of a doctoral research project might respond to each objective. Therefore by adapting the work of Cleaver and Franks, I set out these different demands in the context of my research in terms of scope, focus, and type of data and results, as shown in Table 21.1.

Before beginning my fieldwork I had recognized the need to consider how my research could respond to these three areas, but had agreed with WaterAid that the forms of the contributions to policy and practice could be discussed and clarified as the fieldwork progressed. In fact, I had already undertaken fieldwork on a related topic with WaterAid for my master's degree dissertation and I was able to use time at the start of my doctoral research fieldwork to finalize two case-study reports based on this previous research. This acted as a way of presenting some research for policy purposes, primarily as additional evidence for staff of WaterAid in the UK (as part of moves in the organization to focus more on the sustainability of its interventions) and Mali (to help demonstrate the work of WaterAid's Regional Learning Center in promoting research and learning for decentralized water and sanitation services in West Africa). This work also helped me demonstrate to WaterAid staff in Mali my commitment to the policy relevance of the ongoing doctoral research. These reports served as examples of the types of outputs that could be produced even if I later felt that the collaborative process itself was as important as the tangible 'products' that emerged.

Table 21.1 Research with WaterAid in Mali

	Research objective		
	Generating knowledge	Informing policy	Guiding practice
Definition of scope			
Cleaver and Franks 2008	By researchers	By policy makers	By users
Present research	By the supervisory team in broad terms, later refined by the researcher	By the researcher and WaterAid's policy and advocacy team (trying to influence policy makers) and programs team (trying to improve interventions)	
Focus			
Cleaver and Franks 2008	Improved understanding of the world around us	Evidence of outcomes	Guidance for interventions
Present research	Improved understanding of how and why the costs of water provision are shared between different actors	Evidence of the costs (inputs) and effectiveness (outputs) of WaterAid's approach to working with local governments	Guidance to improve the effectiveness of the work of WaterAid's partners with communities and local governments
Type of data and presentation of results			
Cleaver and Franks 2008	Intensive or extensive empirical research with findings generalized to theoretical propositions and to raising further questions; uncertainty accommodated	Generalized, focus on 'success stories' and 'best practices' with lessons for 'scaling up' and 'scaling out'; certainty of linkages (inputs and outputs) required	Specific and localized, often presented as tools or checklists
Present research	Extensive qualitative and quantitative research, relating the findings to academic theory on financing decentralized public services and the role of NGOs	Lessons which could help promote the adoption of WaterAid's approach ('scaling up') by other actors; some demand for 'success stories' of 'best practice' to share within WaterAid's partners	Tools for monitoring the costs of water provision and analysing users' willingness and ability to pay which could be used by WaterAid's partners

Source: Adapted from Cleaver and Franks (2008).

Starting out: a typical fieldwork experience?

My focus for the first period in Mali was perhaps a fairly typical experience of doing doctoral fieldwork in the Global South. After spending several months in language training (in French and Bambara), I also engaged my own interpreter and travelled with him to the villages where I planned to do in-depth qualitative research of how payments for access to drinking water were organized in different communities. This was a little removed from the day-to-day practical activities and immediate policy requirements of WaterAid and its partners, as I was looking for examples of 'interesting practice' at community levels rather than 'best practices' to be replicated.

During this stage of the fieldwork I tried to follow the advice of Mercer (2006) on working with NGOs by attempting to establish and maintain an independent identity for myself so that I was not seen by research participants (water users and other stakeholders such as local government staff) as a representative of WaterAid or its partners. This challenge, and the related concern of avoiding raising the expectations of research participants that immediate action might be taken as a result of research outcomes, were the ethical issues that I had considered most intensively before beginning fieldwork. On reflection these were difficult to address; despite my trying to explain my position as a researcher, I do not think that most research participants made the distinction between this role and that of others who were actually working for WaterAid. However, I also think that this issue was less important than the wider ethical commitment of trying to ensure that the research had some broader benefit to WaterAid and the populations it works with.

Although I saw these first few months of fieldwork as a time to focus on the main data collection required to satisfy the academic part of my doctorate, they were also an opportunity to maintain contact, build up relationships, and develop further ideas with WaterAid staff, as well as learning from participating in other WaterAid workshops and events.

Part way through: a changing role

After I had spent the early period attempting to establish my position as an independent researcher, my relationship with WaterAid changed just over halfway through the fieldwork. As part of broader moves in WaterAid internationally to address the challenge of the sustainability of water and sanitation interventions, WaterAid organized a regional workshop in Liberia for representatives of its different country programs in West Africa to discuss the implementation of its new Sustainability Framework (WaterAid 2011) in relation to rural water services. I was invited to attend because my research was addressing a key aspect of sustainability. Because of this, I became – as one of the organizers put it – a 'de facto member of the WaterAid Mali team' for the purposes of discussions about how to use the framework to address the challenge of sustainability in Mali. The results of the workshop included each country program drafting an action plan for the research required to guide WaterAid's practice and inform national policy regarding sustainability in their country of work.

In hindsight, I realize that I had been practicing what Eyben (2010) calls 'planned opportunism'; I had known that there was a growing movement within WaterAid internationally to more explicitly address the problem of sustainability in the sector, and the 'launch' of the Sustainability Framework in the West Africa region was a moment when I saw that my support could potentially contribute to some of these changes in the context of WaterAid's work in Mali. Owing to my presence at the workshop, the relevance and flexibility of my ongoing fieldwork, and the temporarily reduced capacity of WaterAid's programs and policy teams in Mali at the time[2] (due to changes in staffing structure including secondments and ongoing recruitment), I became the joint lead for the proposed research on sustainability in Mali. My role was therefore somewhere between technical consultant and research manager. At the same time, I was still hoping from my perspective as a doctoral student that this research could contribute data which I would be able to use for the academic knowledge required by my thesis (see also Chopra in this volume on being both an academic and an activist).

Although I thought that this role could help fulfil part of how I saw my ethical commitment to WaterAid, I was aware at the time that there were potential ethical pitfalls involved. I certainly wanted to avoid influencing the research direction too much so that it became a vehicle for gathering additional data which would serve only my academic work and not the requirements of WaterAid. Related to this concern was the possibility that my involvement would reduce ownership of the process by WaterAid and its partners, so that any potential changes to policy or practice suggested by the research would be less likely to be adopted. A final possibility was that the research went too far in the other direction to become a practice-oriented project with insufficient methodological rigor to be used as part of my doctorate. Bell and Read (1998) specifically caution against this as part of their advice to students working on collaborative projects.

I tried to mitigate these risks by working with WaterAid staff to develop an iterative process for the research, where the exact themes, questions, and approaches were developed through a series of workshops with representatives of WaterAid's partner organizations. The fieldwork was carried out between the workshops by WaterAid's partners using tools that I had drafted but which had been discussed and validated in the workshops. I also took the lead on the initial data analysis, but these results were discussed as much as possible so that the partners could draw out the implications for their own activities. I conducted short follow-up visits to some areas where the teams had identified potentially interesting findings in order to do further qualitative research which could contribute to my thesis. In this way the research that was done primarily for policy and practice could be extended into research for academic knowledge as well.

Although at some points during the research activities I felt over-involved in the details of the process (rather than simply being an advisor on research methods), I agree with Carr (writing in Simon *et al.* 2011) that there are wider benefits to academic development geographers (and others doing research on international development) of working *in* a development organization:

> without an understanding of mundane bureaucratic moments such as budgeting, contracting, and monitoring and evaluation it is simply imposs-ible to understand why agencies do what they do, or reliably to identify points of intervention that might change practice in the world.
>
> (Carr in Simon *et al.* 2011: 2797)

In my case the benefits came from working closely enough with WaterAid's partners to understand how the possible practical lessons for WaterAid emerging from the research might be enabled or constrained by the organization's existing annual cycle of planning, budgeting, monitoring, and reporting. For example, the partners agreed to develop approaches for more closely monitoring the function-ality of all the water points in their areas of intervention. However, the figures they had to report to WaterAid were the numbers of new water points con-structed (or old water points fully rehabilitated) in the relevant reporting period. Therefore, given limited time and resources there was less incentive for them to undertake the more detailed monitoring of functionality. At the time of writing, this reporting process was due to change across all WaterAid's country programs to include consideration of the actual operation of water points up to ten years after their installation, which would address this issue.

From the partners' perspective they saw two ways in which contributions from an academic perspective could benefit their own work. The first of these was in relating their practice to wider academic and sector debates through raising questions and suggesting ideas. The second was in the support of devel-oping data collection and analysis tools. Both of these possibilities had been identified as potential benefits at the start of the collaboration but thinking more in terms of the outputs (policy reports, tools) rather than the actual process involved. On reflection, I argue that the process was just as important as a means for everyone involved to learn from each other in ways that might provoke further critical reflection about these ideas in future work.

Finally: practitioner, researcher, or both?

As well as the benefits that I have argued above I also have to acknowledge that my willingness to work with WaterAid in this more direct research management role was related to my own background as more of a practitioner than researcher. Before starting my postgraduate studies I had been working as a water and san-itation engineer in the implementation of infrastructure projects for an NGO in Kyrgyzstan and had previously managed a small NGO in the UK. I enjoyed the hands-on management aspects of these roles and was eager to take the oppor-tunity to include more of this type of role within my doctoral project when the chance arose, both for my own personal satisfaction and for my professional development.

However, I was also conscious of the need to remain reflective in this 'development manager' part of my overall role. As Abbott *et al.* (2007) and Wilson (2006) argue, development practitioners should be aware of the criticism of them as 'technocrats' legitimizing a particular form of Western development (see Kothari 2005 for an example of this critique), and should seek ways of promoting learning together with those they are hoping to benefit. This was what I tried to support in the research process with WaterAid and those working for its partner organizations, although its extent was limited; our focus was on learning within this group of NGO and local government staff, with much less involvement of the actual water users themselves.

Finally, the period immediately after returning from fieldwork was important to me for further reflection and additional feedback on how I had tried to balance the different issues discussed here. I gave a presentation at WaterAid's London office on how my academic fieldwork and the other research activities had developed in Mali, which allowed me to discuss these issues of balancing objectives and ownership with others who were experienced in research with both academia and NGOs. Following this I was also invited to present at a larger learning event for organizations working in the rural water sector, which fed in to a wider debate about how different groups (academics, practitioners, donors) can contribute to learning in the sector. Both these occasions were useful opportunities to reflect on the ethics of my engagement with WaterAid during fieldwork and highlighted the importance of discussing these sorts of issues with others.

The potential of collaborative research

For me, the key ethical challenge that I encountered as a postgraduate student doing field research with an NGO was how to balance the demands that the research must contribute to knowledge, policy, and practice. There is no simple way to negotiate this challenge, but I would recommend that researchers who are involved in collaborative projects try to identify how their research might respond to these differing objectives. A method similar to the one that I used may help to suggest where there might be mutual benefits or potential conflicts between the different elements. I would recommend doing this with the partner organization early in the research process, especially as the project moves from a broad outline to more specific objectives. However, it is important to remember that the balance between the different elements may change over time, requiring a continual reflection and renegotiation of position. An acknowledgement of 'planned opportunism' (Eyben 2010) may also help researchers to prepare for and react to shifting priorities and opportunities within their partner organization.

My experience shows that collaborative studentships such as the CASE scheme can be a good way for doctoral researchers to help bridge the demands of research for knowledge, policy, and practice. However, this may require closer engagement with the partner organization than simply adopting the basic 'characteristics of a good employee' (meeting deadlines, respecting the partner organization's interests) proposed for CASE students by Bell and Read (1998: 27). I argue that there may be times when the doctoral student can go beyond this and temporarily act more like an actual employee of the collaborating partner for the benefit of both parties. This is linked to my view that thinking about *process* is as important as the eventual *products* of collaborative doctoral research, especially for the parts of the research seeking to improve practice. This was reflected in my own learning about the approaches and internal workings of WaterAid's partners and in their experience of developing collaborative research with a doctoral student, a process which may help inform analysis that they undertake in their future work.

Acknowledgments

I would like to thank all my research participants and the staff of WaterAid and its partners in Mali. This work was supported by the Economic and Social Research Council grant ES/G030243/1. Helpful comments on this chapter were provided by Vandana Desai, Alex Loftus, and Tom Slaymaker.

Recommended reading

Cleaver, F. and Franks, T. (2008) 'Distilling or diluting? Negotiating the water research-policy interface,' *Water Alternatives*, 1 (1): 157–176.

A discussion by academics who have also worked as practitioners and consultants on the challenges of balancing research for knowledge, policy and practice.

Simon, D., Sidaway, J. D., Yeboah, I. E. A., O'Reilly, K. and Carr, E. R. (2011) 'Geographers and/in development,' *Environment and Planning A*, 43 (12): 2788–2800.

A set of short papers based on a panel discussion at the annual conference of the Association of American Geographers in 2011 which provide different perspectives on the role of academic geographers in development. The opposing positions discussed by Kathleen O'Reilly and Ed Carr are particularly thought-provoking in relation to the issue of engaging with development agencies, and are also relevant to work with NGOs.

Notes

1 Despite the name, CASE studentships are for research in the social sciences.
2 As Carr writes in his contribution to Simon *et al.* (2011: 2797), 'in understaffed agencies, as most are, it is startling the number of events and outcomes that are influenced by the simple issue of who has time to look over the documents or attend the meeting in question.'

References

Abbott, D., Brown, S. and Wilson, G. (2007) 'Development management as reflective practice,' *Journal of International Development*, 19 (2): 187–203.

Bell, E. and Read, C. (1998) *On the CASE: Advice for collaborative studentships*, Swindon: Economic and Social Research Council.

Cleaver, F. and Franks, T. (2008) 'Distilling or diluting? Negotiating the water research-policy interface,' *Water Alternatives*, 1 (1): 157–176.

Eyben, R. (2010) 'Hiding relations: The irony of "effective aid",' *European Journal of Development Research*, 22: 382–397.

Kothari, U. (2005) 'Authority and expertise: The professionalisation of international development and the ordering of dissent,' *Antipode*, 37 (3): 425–446.

Lockwood, H. and Smits, S. (2011) *Supporting rural water supply: Moving towards a service delivery approach*, Rugby, UK: Practical Action Publishing.

Mercer, C. (2006) 'Working with partners: NGOs and CBOs,' in V. Desai and R. Potter (eds), *Doing development research*, London: Sage.

Simon, D., Sidaway, J. D., Yeboah, I. E. A., O'Reilly, K. and Carr, E. R. (2011) 'Geographers and/in development,' *Environment and Planning A*, 43 (12): 2788–2800.

WaterAid (2011) *Sustainability framework*, London: WaterAid.

Wilson, G. (2006) 'Beyond the technocrat? The professional expert in development practice,' *Development and Change*, 37 (3): 501–523.

22 'So what kind of student are you?' The ethics of 'giving back' to research participants

Experiences from fieldwork in the community forests of Nepal

Samantha Staddon

It was the very last focus group of my third and final phase of fieldwork. I'd had a wonderful time doing my research in rural Nepal but was ready to go home to my husband. Just as I was beginning to wind down the meeting, one of the participants – a particularly eloquent and quietly confident man – asked me a question which went straight to the heart of the ethical dilemmas that I'd been struggling with throughout my fieldwork. He described two other PhD students who had worked in the area in previous years. One, a German geographer, maintained close connections to the community and had helped raise funds to establish a school (I had been told of this man many times during my fieldwork). The other, a French anthropologist, had never been heard of again. This gentle man then asked me directly: 'So what kind of student are you?' For just about the only time during my fieldwork, I was glad of working through a translator, as it afforded me some extra time to digest the question before I attempted to address it.

We get a certificate, some letters after our name, and a step up on the career ladder out of our doctoral research, but what about the people and communities involved in our fieldwork? Along with academic qualifications, we also gain from the fieldwork situation in terms of experiencing other cultures which would be inaccessible to us were it not for the purpose of doing research (Chacko 2004). Contemplating her own fieldwork experiences, feminist geographer Cindi Katz writes that 'I have learned and built a career on each of these undertakings. Their benefits to participants not withstanding, these field projects all have probably been more beneficial to me than to them' (Katz 1994: 71–72).

Many of us, nonetheless, question what it is that we *can* 'give back' to those people who have spent their valuable time being interviewed or participating in focus groups and the like. While most of us who work in the field of development studies hope our research can help in some small way in the fight against global social injustice and inequality, we often wish to give something directly to the participants involved as well. Maybe out of guilt and political imperative we wish to re-balance the burden of giving and receiving. But 'giving back' to research participants in the Global South is often fraught less by resource constraints and more by ethical dilemmas.

Of course, some of these dilemmas can be addressed through employing participatory research approaches in order explicitly to bring in the priorities and desires of those with whom we are conducting research (Kesby 2005; Van Blerk and Ansell 2007; also Fagerholm in this volume). We could also combine academia and activism (see Chopra in this volume), or engage in policy-relevant research (Pain 2003; Autonomous Geographies Collective 2010; also Jones in this volume). No doubt elements of these approaches creep in to most research; however, this chapter focuses on research that employs 'conventional' methods and approaches, such as interviews, focus groups, surveys, and/or participant observation. While not necessarily directly aiming to empower research participants, work based on such methods nonetheless has an ethical duty to consider how it may 'give back' to those involved.

Issues of 'giving back' are closely related to those of positionality and reflexivity in 'the field,' and while many excellent discussions of these issues exist (Katz 1994; Kobayashi 1994; Nast 1994; Sultana 2007; plus many of the other chapters in this book), few relate specifically to 'giving back.' This chapter aims to take on board such writings in its exploration of the ethical dilemmas of 'giving back' to research participants – although recognizing that 'we are always already in the field' (Katz 1994: 67), and thus 'giving back' can extend beyond our time in 'the field.'

In the next section I shall outline the nature of my doctoral fieldwork because this will allude to some of the choices I faced in terms of collaborations, accommodation, and modes of fieldwork. I then go on to discuss the ethical dimensions of such choices, paying particular attention to the ways in which I chose to 'give back' to the communities involved in my fieldwork. I look at two key aspects: *what* is it appropriate to 'give back' and *who* is it appropriate to 'give back' to. I shall also reveal how I replied to the question, as articulated by the man described in my account at the start of this chapter, of what kind of student I was.

I conducted all my fieldwork with my Nepali research partner, Mr Shyam Shrestha, and this chapter gives emphasis to his presence and positionality: something that is often overlooked in accounts of fieldwork, which tend not to look beyond the individual academic (but see Twyman *et al.* 1999 and Leck in this volume). Although this is a single-authored chapter, I use the word 'we' when referring to fieldwork conducted by myself and Shyam. In other instances I use the first person and this refers to my own thoughts and reflections on the fieldwork process.

Fieldwork in the 'community forests' of Nepal

My doctoral fieldwork took me to Nepal, where I worked with two Community Forest User Groups (CFUG) in the Middle Hill's district of Ramechhap: Golmatar Paleko CFUG and Burke CFUG. My research aimed to investigate how subsistence farmers keep track of changes in the local forests on which they rely for agricultural and domestic inputs, i.e., to understand individuals' qualitative observations of environmental change (Staddon 2012). The research also explored the impacts of 'participatory monitoring' projects in each user group, in which community members worked alongside forest technicians to formally monitor experimental plots in the forest using quantitative techniques. I hoped to understand the dynamics of these informal and formal methods of monitoring (based on indigenous and scientific knowledge respectively) and to see how they may contribute to sustainable resource use, as well as how they may affect nature–society relationships.

I conducted my research in collaboration with the Nepal Swiss Community Forest Project (NSCFP), which has offices in Ramechhap and two neighboring districts, as well as a head office in Kathmandu. As a bilateral aid program, NSCFP has supported the government-led community forestry program in these districts for over 20 years. It was NSCFP district staff who introduced me to the two communities with which I worked and it was NSCFP-initiated participatory monitoring projects that I was investigating. The NSCFP staff I worked with were highly engaged and motivated, and importantly were open to the sort of critical questioning that my research involved.

Throughout my fieldwork, I worked with my Nepali research partner, Shyam, who grew up in the district in which we worked but went on to train to postgraduate level in forestry in Kathmandu. He introduced me to Nepal's community forestry program – in fact to everything Nepali – and the research would not have been the same without his considered and intelligent input. Shyam provided all translation in the field and ensured that all fieldwork was conducted in a culturally sensitive manner (i.e., he sat around chatting to other men for long periods before anything started 'for real'!). While in the field we stayed in the NSCFP district office accommodation and in a small hostel frequented by other local NGO staff, respectively an hour and half-an-hour's walk from the two communities we worked with. Local families provided us with food during the day, for which we paid on a daily basis.

My fieldwork included semi-structured interviews, a household survey, 'harvesting trips' (where we accompanied people in the forest), focus groups, and participant observation. Research participants included those who were involved in the participatory monitoring projects we were studying and other members of the communities who had not directly been involved. Of course, our presence in the two villages meant that many, if not all, community members were aware of our work, and there was much informal contact with the wider community. While we did not talk with every member of a household, our discussions with one member necessarily had an effect on others, e.g., additional chores for them while the member we spoke to was busy with us. This is important to note because although research may target particular individuals as research participants, it undoubtedly has an effect on the wider community.

In order to fit in with family commitments I decided to conduct the fieldwork over three separate visits, talking with new people each time, but also spending time with those we had already spoken to. This arrangement actually worked well because, as my supervisor told me, 'something grows whilst you're away' and it was nice to return for each subsequent trip, as if meeting up again with old friends. At the end of my second visit, my husband came to visit me and we spent time with both of the communities with which I was working. In total I spent eight months with these communities and it was an experience that I will never forget, being far, far more than an academic training exercise.

What to 'give back'?

There are of course many things that we can give to research participants; the problem is that there seem to be ethical dilemmas surrounding all of them! We can of course give financial reimbursement in exchange for the time that respondents give us during interviews and other activities. While this potentially straightforward exchange recognizes the opportunities foregone while respondents engage with us (particularly in economically poor countries) and the intellectual contributions they have to offer (Garnett *et al.* 2009), some researchers highlight issues of postcolonial dependency and the 'standard' set by researchers who introduce payments for responses, which others are then required to follow (Sultana 2007).

In accordance with such debate, during our initial fieldwork in Nepal we chose not to make payments for interviews; however, as time went on and as we were asking some of the same participants to speak to us for a second or third time, and as our fieldwork coincided with the busy agricultural time of the year, we chose to give payments equivalent to local wage labor in exchange for attendance at focus groups. As we only gave people the money at the end of the focus group, we believe it did not affect their willingness to participate or the responses given. The decision to make financial payments is of course also a practical issue, as many students do not have sufficient funds to make this choice a consideration. The payments I made came from my research fund and were accepted graciously by participants, although we did not dwell on the issue, which felt to me like some sort of taboo, or at least carried the risk of embarrassment on both parts.

As an alternative to money, some researchers choose to give photos to research participants (Chacko 2004). With digital cameras this can be an easy option, particularly if researchers leave the field for a period of time and can return having had photos processed. Photos can serve to offer a connection between different phases of fieldwork and can be viewed as a 'neutral' gift, but when we tried to do this in Nepal it somewhat backfired. I had made sure I returned to the field for my second visit laden with photos of those people I spoke to during my first visit. Tracking these people down again proved somewhat problematic, but the main issue to emerge was that people (often children, but not always) in the background of photos also wanted copies and felt sidelined that they hadn't received any. Those that weren't in the photos at all also wanted me to take their photo and give it to them during my next visit: although this was possible, the logistics became increasingly difficult, particularly not knowing names or addresses of individuals in the photos. I am sure there is a place for giving photos, but it is not as simple as it might first appear.

Along with a demand for (more) photos, research participants may request other things, thereby making our dilemmas over what to give them somewhat superfluous. In our fieldwork in Nepal we were frequently asked for technical advice as to how best to manage the community forests of the two CFUGs. These requests reflected the commonly-held perception that CFUG users lacked the technical skills needed to manage their forests sustainably, and rather that government and development professionals alone held the scientific knowledge necessary for this (Nightingale 2005). Although our research aimed explicitly to counter these views – for example by promoting recognition of local ecological knowledge and forms of environmental monitoring – we found ourselves *a part of* the discourse we were trying to work against. Even after asking respondents questions directly related to their local expertise, we were still asked for our seemingly 'better' understanding of the situation and for our practical recommendations.

Such ethical dilemmas have been noted by others, for example Sundberg (2004), who relays how she was asked by the indigenous women's cooperatives with which she worked to provide technical advice. Given her position of assumed authority, Sundberg saw this as reinforcing the performance of *white woman helping* and *indigenous woman in need of help* – something which she was explicitly trying to work against. Others emphasize the need to respect people's requests and that in so doing you can provide information that is otherwise inaccessible to them (Chacko 2004).

The latter is the line of thinking we adopted in our fieldwork, and we responded to people's requests as best we could – acutely aware that government forestry officials were under-resourced and were few and far between, meaning their engagement with local CFUGs was minimal. While offering advice, we made sure that we emphasized the CFUG members' own knowledge and agency in the management of their forests. When I say 'we' here, I actually mean Shyam; as a trained forester he was able to provide the sort of technical advice that was being sought (something which I could not provide). This highlights the importance of collaborations and the role of research partners and others in the outcomes of our research. Had I been working with 'merely' an interpreter, who had no knowledge of the community forestry program that we were studying, the possibilities to 'give back' to the research participants – on the terms that *they* wished for – would have been severely restricted (see also Leck in this volume, who worked with her research assistant to explain to participants more about climate change).

Almost as an antithesis to the technical, formulaic forestry advice requested, the forms of daily interactions with research participants and other community members and the building of relationships with people were an arena in which I also felt we could 'give something back,' albeit in a far more subtle way. Sultana (2007) discusses the 'everyday acts' of fieldwork, including the decisions of what to eat, where to sit within a person's house, what to wear, and how to address people. She concludes that 'such little actions, however mundane, are not insignificant I believe, and speak to the embodied situatedness of me as the researcher that I had to constantly keep in mind' (Sultana 2007: 379). In our fieldwork, we too attempted to ensure that our actions demonstrated respect and mindfulness. Our respect toward children was noted by some members of the community, for example, who felt that this was a good thing, despite the fact that it is not a part of local culture.

Some academics talk of the tension between private and research personas when in the field (Chacko 2004), while others interestingly discuss personality as 'the new positionality' (Moser 2008). Issues of 'embodied situatedness' in the field (c.f. Sultana 2007) call for a holistic view of fieldwork that recognizes that every action we take while in the field has repercussions and that we cannot break these down into personal *or* professional; rather, I believe, these are bridged by our personalities, which can have a huge impact on our research (see also Smith in this volume).

Sultana (2007) also discusses the wish to become friends with some of her research respondents, something which they apparently found a strange concept, given the class divide between them. One family that I became particularly friendly with during our fieldwork in Nepal were of a local *dalit* caste, and while they seemed as happy with this relationship as I was, I was conscious of the potential opinion of others, particularly higher castes within the community (although I was never made aware of such feelings, if indeed they did exist). I did nothing about this, however, and actively maintained my friendship with the family.

Most of us hope that if we effectively disseminating our research, possibly to a variety of audiences (Staehili 1994), its findings and message will in some diffuse way benefit the research participants involved and thousands of others like them throughout the Global South. The relevance of academic papers to research participants, particularly those whose main desires may be for material improvements to their lives, is potentially dubious. More fundamentally, however, the ability and morality of academics to speak of/for members of the Global South has been thoroughly critiqued. This 'crisis of representation,' as it is referred to, centers on the ethico-political implications of Western academics (or the elite global professional class) speaking for marginalized groups and the subaltern in the Global South (Katz 1994; Nast 1994; Kapoor 2004; also Gent in this volume). As stated in Kapoor (2004: 631–632):

> Though the speaker may be trying to materially improve the situation of some lesser-privileged group, the effects of her discourse is to reinforce racist, imperialist conceptions and perhaps also to further silence the lesser-privileged group's own ability to speak and be heard.

Kobayashi (1994) blames the crisis on naive assumptions about researchers' ability to contribute to others; however Sultana (2007) reminds us that to give up on fieldwork as a result will rule out opportunities to work against global inequalities.

While some researchers may assume that disseminating findings has little to do with fieldwork, others argue that the politics of representation is closely linked to the politics of fieldwork (Nast 1994). Just as our ideas of how to analyze, structure, and present our findings begin during our time in the field, the potentially unequal power relations between researcher and researched suffuse our work, meaning fieldwork can never be separated from the representational texts that emerge. Efforts at research dissemination are therefore fraught with their own ethical dilemmas; however, having chosen to conduct fieldwork in the Global South, I chose in my work to accept the political imperative to promote an awareness of the politics and social inequalities I witnessed. By drawing epistemologically, methodologically, and politically from critical and feminist social sciences but by writing explicitly for a natural science audience, I hope my work will cross disciplines and bring issues of knowledge, power, and politics to the attention of ecologists and conservationists.

The flip side of disseminating findings and arguments far and wide is the necessity to 'feed back' to those actually involved in the research. Whether through formal feedback sessions at the end of a research project, or continual, informal chatting to respondents to let them know how the work is progressing, it has been suggested that feedback is an 'obligation' when conducting fieldwork in the Global South (Van Blerk and Ansell 2007). Feedback in our research in Nepal took a variety of forms, both verbal and written; however, dilemmas existed, particularly in the latter. Providing a written progress report or final report to a community in which many (particularly women) are illiterate is problematic, as is ensuring access to the reports for all (see also Fagerholm in this volume).

During our fieldwork we noted such problems in relation to the 'Operational Plan' of the CFUGs, which is a legal document setting out the management plans for the community forest, written in consultation with the local District Forest Office. Given that many in Nepal's CFUGs cannot read these plans, others in the group have to read them for them, which gives these literate members particular power over others. While this power can be challenged, for example by illiterate members questioning how they can trust what is read to them given that they cannot verify it (Nightingale 2010), similar issues may emerge in relation to documents we provided to the CFUGs as feedback of our research. As I publish papers based on my research, I hope to send copies to the CFUGs involved; however, given issues of literacy and more importantly (at least symbolically) access, the extent to which this represents an appropriate form of feedback, or thanks, is questionable. The heterogeneity inherent in many communities (for example based on literacy) raises another important consideration when negotiating the dilemma of 'giving back' to communities involved in fieldwork, that of who in the community in which we work do we wish to 'give back' to – which is the focus of the next section.

Who to 'give back' to?

Ideally we'd love to 'give back' to *all* those in the communities in which we work, but most of us recognize that communities are not homogenous and that our efforts are necessarily restricted and importantly involve choices. Deciding who to give to – and who *not* to give to – obviously depends on what it is we are giving; the choice of who to give money to and who to give photos to will be different. However, in both cases it is important to recognize the partiality of our acts and that while we may please some people or parts of the community, it is unlikely that we will satisfy them all. Despite our best intentions, this may inadvertently contribute to existing divisions within the community.

As revealed above, some of the things we 'gave back' during our fieldwork in Nepal were aimed at particular individuals (e.g., money to respondents in particular focus groups) while others were directed more generally at the wider community (e.g., respect in our relationships with all people or our provision of technical advice and the feedback of results). Also noted above is the potential problem of elite capture, or preferentially benefiting those who were literate and could therefore read our written reports and recommendations. This of course was a real possibility in relation to our own work in Nepal.

Not mentioned in the previous section was my decision to make a contribution (again, made from my research funds) to the funds of the two CFUGs with which we worked. This was made at the end of the fieldwork and in recognition of the central role the CFUG played; I could have donated money to the local school, for example, but the research had nothing to do with this and thus I would not have felt particularly comfortable doing that. The funds of Nepal's community forestry groups are intended to be used partly for forest management activities and partly for community development and help for the 'identified poor' within communities. A donation to the fund could therefore be used in a variety of environmentally and socially sound ways, decided upon by the community themselves. A cursory read of the community forestry literature and our own fieldwork experiences, however, tell of the widespread problem of elite capture in CFUGs and the lack of transparency in the allocation of funds. Thus by giving to the fund, we could have been contributing to such issues. The decision to make a donation was difficult, but I felt that it was an appropriate way to 'give back' in the immediate term (sorting out problems of corruption are long-term issues) and one which was in line with the aims of the research.

As mentioned above, the provision of advice, information, or recommendations may (potentially) benefit the community at large (if not each and every member), and this was another approach we took in our work in Nepal. As researchers we also hope that the dissemination of our work may make it applicable and relevant to communities *other than/in addition to* those we directly work with, i.e., that it might 'make a difference' on a far larger scale. Despite the politics and ethics of representation, this will be an ongoing way in which I hope my research can be of some benefit.

I have discussed the importance of collaborations above (with individuals and with organizations), and while not strictly a part of the communities our work is directly associated with, these too can be important avenues for our 'giving.' Although deserving of an entirely separate discussion, ensuring fair and adequate resources, training, and terms of employment for our research partners, assistants, and interpreters is a fundamental ethical tenant of fieldwork. Similarly, ensuring open and respectful communications with collaborating organizations should be the priority of all researchers. In my research, maintaining discussions with and links to both Shyam and NSCFP has been important since my return to the UK from Nepal. This has served not only to enrich the research that emerges from the project, but also to enable ongoing communication with and feedback to the two communities we worked with – something that would be basically impossible if relying on postal routes. While I do not plan to return to Nepal to conduct further research in the near future, these connections may potentially prove productive in the future for both parties.

Increasingly, I have come to recognize the importance of using our privileged positions in Western academic institutes for the benefit of friends and colleagues in the Global South, e.g., in accessing journal articles and passing on other scholarly activities and advice. Sometimes I have offered this help and sometimes it has been requested; however, it has always been highly appreciated by its recipients, who have such limited access to academic resources and advice in their home countries.

We should ask not only *what* to give back and *who* to 'give back' to, but also *when* to 'give back.' mrs kinpaisby (2008) promotes the idea of 'slow research,' i.e., developing and nurturing relations with research participants/respondents over long time periods (six years is mentioned in relation to research in New Zealand). Of course, master's degree or doctoral students generally have a defined time period in which to start, conduct, and end their research; however, there are many ways to have impacts beyond the relatively small periods of time we may spend in the field. Already mentioned in this chapter is the importance of providing ongoing feedback, dissemination, help, or advice to the communities we work with. Many academics manage to retain links to the places where they carried out doctoral field research throughout their academic career, and this is certainly one way to promote more sustained efforts in 'giving back' to research participants.

So what sort of student was I?

It is clear that as researchers in the field we face an ongoing barrage of ethical choices with regard to how we relate to and 'give back' to those with whom we work. As Sultana (2007: 379) writes, these choices 'are not captured in the "good" ethical guidelines of institutional paperwork, but have to be negotiated and grappled with on a daily basis in the field.' Our decisions as to how to act and how to 'give' depend on our positionalities and those of our fellow workers and on the local social, political, and economic context in which we find ourselves. There are no right or wrong answers, only situations to be negotiated in the best way we see fit. So, how did I view my overall approach to 'giving back' to the communities involved in my research, as elicited by the man in my final focus group?

In replying to his question 'So what sort of student are you?' I started by outlining the ways in which my fieldwork would feed in to my doctoral thesis and the likely outputs of that in terms of reports, papers, and recommendations. I assured him that I would pass on copies of the work produced, so that the CFUG would have some form of evidence of that – even if the contents were not understandable to them (it would be written in English) or entirely relevant. My intention here was that they would know I had stuck to my word, rather than providing something they could directly use. I continued by discussing the ways in which Shyam had been able to offer practical, technical advice with regards to forest management, both verbally in both communities and soon to follow in a written specific report for each (in Nepali). I also spoke of the monetary contribution I was going to make to the CFUG fund – partly as a way to raise awareness of this and thus hopefully a demand for its transparent expenditure.

I then went on to tackle the issue of the school – and what I *wasn't* going to give them. I spoke about the fact that while I could, in theory (if money were no object, and I did try and convey the fact that students weren't that well-off), 'give' them a school, that that would only really lead to a problem of who was going to give a school to the village down the road, or the one beyond that? If they were not 'lucky' enough to receive a researcher, how would they manage? I attempted to highlight the inequalities inherent in such an action, and that this was something I was trying to work against, not contribute to. I pointed out that in fact as a community *they* were the ones who needed to work for their common good – as they were currently demonstrating in their creation of a tea cooperative, to take advantage of opening markets. This was clear evidence of their ability to self-organize and to secure future economic opportunities – something that I said was far more important that a foreigner coming in and merely handing out lots of money (or a school).

And what did the man, and the others present at the focus group, think of my reply? As a whole, they seemed to be quite satisfied: they grasped the (potential) importance of my research, they very much appreciated the direct advice from Shyam, and they murmured quiet approval for the CFUG fund contribution. They nodded in agreement with my analysis that working in community ventures would provide them with a more sustainable future then receiving handouts. I felt I had passed the gauntlet of the question posed to me, and while having to justify myself had made me nervous at the time, I very much appreciated being able to air my concerns openly with those I'd been concerned about.

In less formal settings during my final field visit, I spoke to individuals about my desire to return to the villages at some point in the future with my family. I'd had many discussions, particularly with women, about my lack of children given my age (33) and the fact that I'd already been married for five years (see also the chapters by Godbole and by Lunn and Moscuzza in this volume), so I promised to bring the two girls I told them I hoped I'd have to visit them in the future. Funnily enough, writing this four years later, I now have two girls, and that makes me feel even more duty-bound to go back and see them and tell them the subject of our conversations had come true.

It is such personal stories that will stick with me from my fieldwork, and I hope one day to be able to demonstrate to the villagers that I have gone on remembering them, despite the time and space that separates us. Whether I will find that they care about the fact that I have continued to think of them, should I return there, remains to be seen. Maybe the jury is still out on what kind of student they think I am, but I'm hoping that I've got another few years in which to prove my worth – and gather the funds for a family trip back there!

Negotiating the ethical dilemmas of 'giving back'

This chapter has probably raised more questions and issues than it has provided 'answers' or 'solutions' to, in line with most of the chapters in this volume. In negotiating the ethical dilemmas involved in 'giving back' to research participants, we must work with local contexts that are constantly evolving. No ethical guidelines can tell us what to expect or what to do when a situation arises. Issues of 'giving back' relate closely to those of positionality, reflexivity, and representation, and thus a reading of those literatures will provide a sound basis upon which to base our negotiations. Being aware that dilemmas will exist is an important first step to negotiating them as they arise.

I find the following two quotes instructive: 'There needs to be a critical and reflexive questioning of what the research/researcher hopes to accomplish, why a particular area was chosen, and for whom are we working' (Nast 1994: 58), and 'we necessarily, through our involvement in the landscapes of those with whom we work, contribute to the production, reproduction, and change of a particular place' (ibid.: 60). *We* choose where our fieldwork takes place (and generally *not* the communities we work with), and intentionally or otherwise, *we will* have an impact on that site. Ethics, morality, and politics demand that we attempt to make our 'contribution' to the communities with which we work as positive as possible – being crucially aware that what is positive for some in the community may not be for others.

This chapter has discussed the ethical dilemmas of 'giving back' to research participants in the Global South, specifically involving fieldwork based on 'conventional' methods such as interviews and focus groups. Despite not actively promoting participatory or 'empowering' methodologies, such research has an ethical duty to consider how it can 'give back' to those involved – necessitating a consideration of *what* it should 'give back,' *who* it should 'give back' to, and also *when* to 'give back.' It has been noted that 'giving back' is not restricted to – nor does it necessarily start or end in – 'the field,' and that issues of representation bring together the different stages of research. How best to represent those involved in fieldwork is of utmost importance to researchers, both to ensure the validity of work and ethically, as a form of 'giving back.' Little literature exists to help guide us through the dilemmas of 'giving back,' which, given that it is based on positionality and context, may take many forms – tangible and subtle – that require negotiation on a daily basis.

Acknowledgments

My deepest gratitude goes to the communities of Golmater Paleko and Burke Community Forest User Groups, Nepal, for their willingness to talk to me, for their patience with my presence, and some for becoming my friends. I am particularly grateful for the question/challenge which forms the point of entry for this chapter and which enabled me to negotiate my ethical dilemmas in the field. I thank Mr Shyam Shrestha and Dr Andrea Nightingale for making my postgraduate work what it was.

Recommended reading

Kapoor, I. (2004) 'Hyper-self-reflexive development? Spivak on representing the Third World "Other",' *Third World Quarterly*, 25 (4): 627–647.

Drawing on the writings of Spivak, Kapoor discusses the 'ethico-political' implications of researchers' representations of the 'Third World Other' and claims that these constitute a continuing form of imperialism. For me this process of representation starts most definitely in the field; this article is particularly relevant to debates on 'giving back' in terms of the impacts of our research – intended or otherwise.

Sultana, F. (2007) 'Reflexivity, positionality and participatory ethics: Negotiating fieldwork dilemmas in international research,' *ACME*, 6 (3): 374–385.

Despite not once mentioning 'giving back,' the entire article is highly relevant to those contemplating such dilemmas.

Sundberg, J. (2004) 'Identities in the making: Conservation, gender and race in the Maya Biosphere Reserve, Guatemala,' *Gender, Place and Culture*, 11 (1): 43–66.

Examining how discourses and practices of conservation projects are instrumental in mapping ways of life that are gendered and racialized, Sundberg reflects on her own part in this story by analyzing the role of ethnographers in (re)configuring social identities through their fieldwork (pp. 55–57).

Waugh, L. (2003) *Hearing birds fly: A nomadic year in Mongolia*, Abacus, London.

Written by a British-German woman who spent a year teaching in the far west of Mongolia, this book brilliantly captures for me some of the ethical dilemmas involved in living and working with communities in the Global South. Waugh discusses dilemmas such as where to live – or rather *who* to live with – in a small community where ethnic divisions are rife and where nowhere is 'neutral.'

References

Autonomous Geographies Collective (2010) 'Beyond scholarly activism: Making strategic interventions inside and outside the neoliberal university,' *ACME*, 9 (2): 245–275.

Chacko, E. (2004) 'Positionality and praxis: Fieldwork experiences in rural India,' *Singapore Journal of Tropical Geography*, 25 (1): 51–63.

Garnett, S. T., Crowley, G. M., Hunter-Xenie, H., Kozanayi, W., Sithole, B., Palmer, C., Southgate, R. and Zander, K. K. (2009) 'Transformative knowledge transfer through empowering and paying community researchers,' *Biotropica*, 41 (5): 571–577.

Kapoor, I. (2004) 'Hyper-self-reflexive development? Spivak on representing the Third World "Other",' *Third World Quarterly*, 25 (4): 627–647.

Katz, C. (1994) 'Playing the field: Questions of fieldwork in geography,' *Professional Geographer*, 46 (1): 67–72.

Kesby, M. (2005) 'Retheorizing empowerment-through-participation as a performance in space: Beyond tyranny to transformation,' *Signs*, 30 (4): 2037–2065.

Kobayashi, A. (1994) 'Coloring the field: Gender, "race", and the politics of fieldwork,' *Professional Geographer*, 46 (1): 73–80.

Moser, S. (2008) 'Personality: A new positionality?,' *Area*, 40 (3): 383–392.

mrs kinpaisby (2008) 'Taking stock of participatory geographies: Envisioning the communiversity,' *Transactions of the Institute of British Geographers*, 33; 292–299.

Nast, H. J. (1994) 'Women in the field: Critical feminist methodologies and theoretical perspectives', *Professional Geographer*, 46 (1): 54–66.

Nightingale, A. J. (2005) '"The experts taught us all we know": Professionalisation and knowledge in Nepalese community forestry,' *Antipode*, 37 (3): 581–604.

Nightingale, A. J. (2010) 'Bounding difference: Intersectionality and the material production of gender, caste, class and environment in Nepal,' *Geoforum*, 42 (2): 153–162.

Pain, R. (2003) 'Social geography: On action-oriented research,' *Progress in Human Geography*, 27 (5): 649–657.

Staddon, S. C. (2012) *Keeping track of nature: Interdisciplinary insights for participatory ecological monitoring*, unpublished thesis, University of Edinburgh.

Staehili, L. (1994) 'A discussion of "women in the field": The politics of feminist fieldwork,' *Professional Geographer*, 46 (1): 96–102.

Sultana, F. (2007) 'Reflexivity, positionality and participatory ethics: Negotiating fieldwork dilemmas in international research,' *ACME*, 6 (3): 374–385.

Sundberg, J. (2004) 'Identities in the making: Conservation, gender and race in the Maya Biosphere Reserve, Guatemala,' *Gender, Place and Culture*, 11 (1): 43–66.

Twyman, C., Morrison, J. and Sporton, D. (1999) 'The final fifth: Autobiography, reflexivity and interpretation in cross-cultural research,' *Area*, 31 (4): 313–325.

Van Blerk, L. and Ansell, A. (2007) 'Participatory feedback and dissemination *with* and *for* children: Reflections from research with young migrants in southern Africa,' *Children's Geographies*, 5 (3): 313–324.

23 Afterword

Katie Willis

It is a great pleasure to write this afterword to a collection of thought-provoking chapters which demonstrate the vibrancy of research in the Global South by scholars based in Northern institutions, and the insights revealed when reflecting on the research process itself. Nearly 20 years ago when Elsbeth Robson and I were doctoral students in Oxford, conducting research in Nigeria and Mexico respectively, we were the PhD student representatives on the Developing Areas Research Group (DARG) Committee. From our own experiences and discussions with other Geography PhD students at UK universities conducting research in the Global South, we felt that there was a need for students to share their experiences of field research to demonstrate the messiness of research reality, often at odds with the accounts presented in research papers or research methods publications. This led us to convene a session at the 1994 conference of the Institute of British Geographers (as it was then). Contributions to this session were edited into a DARG monograph (Robson and Willis 1994), which was revised and expanded into a second edition (1997).

Elsbeth and I were both strongly influenced by feminist debates regarding positionality and the exercise of power within the research process (see, for example, Harding 1987; Stacey 1988; McDowell 1992). In the early 1990s these debates were key elements of the emerging field of feminist geography, and there was also a growing engagement with arguments coming out of postcolonial theory around Northern researchers conducting fieldwork in the Global South (see, for example, Sidaway 1992 and Madge 1993). What the chapters in the present volume reveal is that discussions about the researcher's position and the politics and ethics of research have become embedded in how many geographers approach their research. As I highlight below, what is particularly welcome is the way assumptions about how power works in research have been problematized, so it is not a simple matter of looking at different dimensions of identity and 'reading off' how these will play out in research practice. There are, however, continuing challenges regarding how issues of power and positionality are discussed, both in PhD theses and in publications. They usually appear in the section or chapter on methodology, and while this is an eminently practical choice, it can lead to an impression that these issues are not part of analysis and writing.

The contained and regimented approach to research politics and ethics is very clearly seen in institutional approaches to ethical approval. A major change in UK university policy since I was a PhD student is how ethical approval has extended from the medical schools to other faculties. This has been driven both by internal decisions within universities and by the requirements of funding bodies, such as the Economic and Social Research Council (ESRC). While research approval in the social sciences has long been a part of North American university life through the institutional review board system, this shift within the UK has been very noticeable, although the requirements still vary a great deal between institutions (Blake 2007).

I would argue that making research ethics procedures a compulsory part of conducting research is welcome, but the challenge is what mechanisms to use. As Sarah Dyer and David Demeritt state in their critique of how medical ethics procedures have been implemented in social sciences, 'Ethical review is designed to pre-empt complaint by imposing a system of external accountability' (2009: 48). A number of authors in this volume have also highlighted the problems with the institutional tendency to have a rigid 'check-box' system, where ethical approval is seen as a one-time process, conducted without the input of research participants and without a recognition of the fluid and contingent nature of ethics. However, at the other extreme is perhaps the impossible position of the researcher as an 'ethical super-human,' as Thomas Aneurin Smith suggests in his chapter.

The chapters in this book demonstrate the diverse ways in which doctoral researchers have negotiated this ethical path and their reflections on the decisions that they have made, sometimes without realizing the ethical implications of these decisions at the time. A particular feature of much research in the Global South, and one which all contributors to this book share, is a commitment to 'making the world better.' In some cases, this is an explicit attempt to link activism and academic research, but in others it is through collaboration with local partners or the sharing of results with policy-makers or civil society organizations (see, for example, the chapters by Chopra, Jones, and Staddon in this volume). This broader ethical commitment frames the choice of research topic, but also the way in which it is conducted.

In the rest of this afterword I want to discuss these themes in more detail. In particular, I will reflect on research as a social process and consider the politics of research not just for those directly involved, but also for those indirectly implicated or affected. Then I will move on to examine the ongoing and open-ended nature of research ethics, concluding with a discussion of recent debates on participatory ethics.

Research as a social process

Simplistic assumptions that researchers from the Global North working in the Global South will always 'hold the upper hand' in terms of power in the research process have been increasingly challenged, and as many chapters in this book demonstrate, as a PhD student you often feel that you are at the mercy of other people's willingness to participate or grant access to communities, field sites, or documents (see also Sultana 2007). This is a common issue in all research regardless of career stage, including in my own experience. During research as a PhD student on women's work in low- and middle-income households in urban Mexico (see, for example, Willis 2000) I had numerous situations where women refused to participate, or where they agreed to participate but their responses to my questions were monosyllabic or very vague. As a more experienced researcher looking at gender and identity issues among Singaporean highly-skilled migrants to China (see, for example, Willis and Yeoh 1999), there were similar challenges, leading to frustrations on my part. However, just because there is no simplistic North-South power dynamic in research does not mean that this is an issue which is no longer worthy of consideration.

Linked to particular constructions of the Global South 'Other' has been an often implicit assumption that low-income research participants will share the perspectives of outside researchers, particularly when there is commitment to 'progressive change.' Associated with this may be a belief that any tensions in the research process, or negative emotions, are the result of failings on the part of the researcher. However, as Thomas Aneurin Smith also highlights in his chapter, assumptions of 'moral and ethical alignment with the researched' are sometimes unfounded. Rather than making homogenizing and potentially patron-izing assumptions about research participants, it is important that the context, the topic, and the individuals concerned are considered, and that researchers recog-nize that as with all social interactions, there are people that we 'click' with more than others. It is also legitimate to feel animosity toward individuals, but whether or how this is expressed is a different question. I have reflected on the role of anger in fieldwork and potential reasons for its exclusion from many fieldwork accounts (Willis 2013; see also Briggs 1970). The need to recognize the context–specific nature of social interactions with individuals is also stressed in the themed issue of *Geoforum* on conducting oppositional research edited by Claudia Hanson Thiem and Morgan Robertson (2010). The papers challenge the monolithic and homogeneous characterization of institutions such as the World Bank and express the need to engage with the diverse personalities and politics of individuals within them.

A focus on responsibilities toward research participants within ethics discus-sions usually fails to engage with the implications for non-participants. This may be of particular importance when participation includes potential benefits such as financial recompense or access to information or social networks which could provide advantages in the future. Debates about the ethics of payment for parti-cipation are wide-ranging (see, for example, the chapters by Fagerholm, Day, and Wang in this volume), but as Dan Hamnett and Deborah Sporton (2012) argue, one aspect which is often excluded is the ethical responsibility of researchers toward non-participants. In her chapter, Samantha Staddon does recognize the need for other community members to 'take up the slack' when participants are involved in the research, so their contribution should be recognized. For Hamnett and Sporton, the decision to provide financial support for community facilities in the Kenyan villages where they run undergraduate and postgraduate fieldtrips was based on a desire to thank all community members. The decision was made following discussions with local community leaders and field center staff but led to tensions with some research participants and also with local guides and interpreters. This tension was not only because it was a change in approach from previous years, but also because other visiting student groups were continuing to give individual participants a small token of gratitude (such as a bar of soap or a bag of sugar) for their involvement. This highlights how research practices and their ethical underpinnings are also implicated in wider networks of academic research and also potentially set up expectations for future researchers (see also Knapp in this volume on implications for future researchers based on her experiences in Bhutan).

One dimension of social relations which rarely comes out in the literature, but which is very apparent from personal experiences and emerges from some of the chapters in this volume, is the relations with our loved ones, whether they accompany us on fieldwork or not (see, for example, the chapters by Le Masson and by Lunn and Moscuzza). The need to acknowledge the fluidity of the boundary between researchers' lives 'at work' and 'outside work' has been an important theme of feminist research. In most of the discussions of research ethics, the focus is on the research participants and the need to 'minimize harm' for them. However, in our lives beyond our narrow identities as 'researchers' we are partners, spouses, parents, children, siblings etc. with affective and material ties with family members who may be affected by our research, particularly if there are long periods of absence overseas. It was only in hindsight that I realized the worry that my PhD fieldwork in Mexico had generated for my family. This was in the days before mobile phones and the internet, so communication relied on the postal system and infrequent calls from public phones. Even with greater connectivity which new technology provides those with appropriate resources in many parts of the world, the potential emotional impacts of overseas fieldwork on family members should be recognized. This certainly does not mean that researchers should never set foot outside their front door, but it does imply that thinking about how and when you keep in touch with loved ones 'at home' during your fieldwork is an important and legitimate topic.

Additionally, the impacts on the researcher of conducting research are also rarely considered. While some institutional ethics forms include mention of the likelihood of the research resulting in 'physical or psychological distress or discomfort,' this is not usually an issue which is debated or prioritized, given the focus of ethics procedures on participants. Researcher health is usually covered, at least in the UK context, through the risk assessment process linked to health and safety legislation (see Tomei in this volume). Research is always going to have stressful and difficult elements; these may be related to academic concerns, research logistics, or broader material or emotional aspects. The accounts in this volume are highly informative in providing insights into the concerns and frustrations which fieldwork can engender. While for most researchers, including doctoral researchers, periods of overseas fieldwork are often life-changing in a positive way, this is not the case for everyone, and even for those for whom the experience was positive, there can be difficulties. These may relate to loneliness or physical illness, or distressing research encounters, particularly if dealing with challenging topics such as domestic violence, child mortality, or HIV/AIDS. It is important that researchers are given appropriate support, including counselling support if suitable, when dealing with such issues. The ethics of care relate not only to research participants and those indirectly affected by our research, but also to ourselves.

This section reflected on debates around positionality and highlighted how simplistic assumptions about power in research encounters need to be challenged as many contributors to this volume have done. Additionally, it has stressed how non-participants can be implicated in research and the ethical dimensions of these relationships. The aim was not to provide a still larger burden of ethical dilemmas with which researchers have to engage, but rather to make visible certain elements of research practice which are often ignored.

Ethical research as an ongoing process

Gill Valentine, in her discussion of ethical dimensions of geographical research, argues that 'the danger is that the rubber stamp of an ethical committee can both bureaucratize ethical reflection and also lull us into forgetting the need to take responsibility for thinking ethically on a day-to-day basis' (2005: 485). As I mentioned in the introduction to this chapter, criticisms of institutional ethics procedures often focus on this limited snapshot of the ethical aspects of a research project. This is predicated on the assumption that the ethical aspects of research are easily identifiable at the start of the research and will not change. As many of the chapters in this book reveal, this is never the case.

In her chapter, Nora Fagerholm provides a useful discussion of the different stages of a participatory GIS process, drawing on the work of Giacomo Rambaldi, Peter Kwaku Kyem, Mike McCall, and Daniel Weiner (2006). While no research follows a simple linear pattern, the identification of planning, data collection, analysis, and dissemination stages is helpful, with each stage bringing up different ethical challenges. When researching internationally there is also a clear moment when you leave 'the field' in terms of physical proximity and, as Kamakshi Perera-Mubarak highlights in her chapter on post-tsunami Sri Lanka in this volume, there are ethical aspects of how you leave the field and deliver on your commitments from a distance.

Another aspect of the research process with ethical implications is funding. Research funding is often very difficult to obtain, but researchers, regardless of their career stage, need to consider the potential conditionalities attached to certain funding sources and should always ensure that participants are aware of the funding source as part of the informed consent process (although see Bryant and Skinner in this volume on notions of informed consent) and that funding sources are acknowledged in publications. A high-profile case where funding sources became the focus of specific disagreements was the México Indígena project involving participatory mapping among indigenous groups in Southern Mexico. This project was funded by the US Army's Foreign Military Studies Office through the American Geographical Society's Bowman Expedition Program (see the commentaries in *Political Geography* by Steinberg *et al.* 2010 for more details).

The ownership and dissemination of research findings is another key aspect of the research process which requires ethical considerations. For the readers of this book who are PhD students, I would strongly recommend that you check your university requirements regarding the publication of your thesis. As part of moves to make doctoral research widely accessible, most UK institutions now require students to submit their theses electronically so they can be published online. This has been a common practice in many other parts of the Global North for some time. Significant care needs to be given to ensuring that potentially sensitive material is not made public. For example, Lisa Ingwall King, a Geography PhD student at Royal Holloway working on ecosystem services in Guyana, collected information about fishing grounds from members of indigenous communities in the Rupununi river basin. She mapped these as part of her analysis regarding the temporal and spatial changes in ecosystem service provisioning. However, the details of these fishing grounds were sensitive and the communities she worked with did not want the specifics to be made available to a wider audience. She therefore ensured that the final version of the thesis did not include the maps containing sensitive information (see also Fagerholm in this volume), although copies of the maps themselves are, of course, available to community members as they own the data. This arrangement was agreed with the community representatives as part of the discussions about the research project. The need to exclude particular pieces of information from any dissemination outputs is also discussed by Andrew Brooks in this volume regarding survival strategies of low-income research participants.

An acknowledgement of the ongoing nature of ethical issues in research and the need to understand ethical practice as a negotiated, social process between researchers and participants has been a particular focus of research adopting a participatory approach, especially participatory action research (PAR). For Caitlin Cahill, Farhana Sultana, and Rachel Pain in their introduction to a special issue of *ACME* on participatory ethics: 'Our conceptualization of a participatory ethics is motivated by a vision for "what could be", and the possibilities of addressing asymmetries of power, privilege, and knowledge production' (2007: 306). This ethical understanding of the nature of research focuses on a form of engaged scholarship and collaboration.

While participatory research approaches have become more common in research in the Global South, there are significant obstacles to fully-fledged PAR projects, particularly for graduate students who have not had previous engagement with the communities and/or organizations they end up working in and with. Working with a community to identify a problem which needs to be tackled and then following through with a participatory methodology, and submitting a thesis within the time-frame required, is decidedly challenging. However, the broad debate about participatory ethics can also inform other kinds of research in the Global South. For example, Megan Blake (2007) discusses the concept of 'negotiated consent' and how the logistics and timing of this differ from the formalized 'informed consent' paperwork which ethics reviews often require. Participants are given information about the research before an interview or other activity, but it is only after the interview that consent is more formally discussed. This allows the participant to make a more informed decision and to be specific about which pieces of information they would like withdrawn, whether they want to be anonymous and so on. Consent can be given in writing, or through audio/visual recording.

Similar approaches have been adopted in community-led research. Project COBRA (Community-Owned Best Practice for Sustainable Resource Adaptive Management is a collaborative project between European and South American university researchers, civil society organizations, and indigenous communities involving participatory mapping, video, and photography, among other approaches, to develop community-owned resource management solutions (see www.projectcobra.org). Discussions about the nature of the research, the methods to be used, the ownership of the outputs, and the dissemination of the results were all completed before the project formally began, but there is an ongoing process of negotiation with individual participants, communities, and organizations as the project unfolds. An openness and willingness to discuss tensions and potential problems has also been important in providing space for negotiations about research ethics (see Mistry and Berardi 2012 for discussion of these issues in earlier projects).

Conclusions

This volume provides important, honest insights into the grounded realities of conducting human and environmental research in the Global South. Across the chapters the clear messages are that there is no one correct way to conduct research 'ethically,' despite what formalized paperwork from ethics review boards suggests, and that everyone has doubts about the decisions they make during research. It is also very likely that we have all done things in our research that in hindsight we would do differently. In this chapter I have been wary of attempting to provide answers to ethical dilemmas that we face in our research as I wanted to stress the contingent, relational nature of ethical praxis. However, as a final point I want to reinforce the importance of sharing concerns and dilemmas about how we conduct our research, what questions we ask, who we work with, and how our research is used. Books such as this one are an important element of these debates, but so are the more informal discussions we have with collaborators and research participants, as well as with colleagues, fellow students, and supervisors.

References

Blake, M. (2007) 'Formality and friendship: Research ethics review and participatory action research,' *ACME*, 6 (3): 411–421.

Briggs, J. (1970) *Never in anger: Portrait of an Eskimo family*, Cambridge: Harvard University Press.

Cahill, C., Sultana, F. and Pain, R. (2007) 'Participatory ethics: Politics, practices, institutions,' *ACME*, 6 (3): 304–318.

Dyer, S. and Demeritt, D. (2009) 'Un-ethical review? Why it is wrong to apply the medical model of research governance to human geography,' *Progress in Human Geography*, 33 (1): 46–64.

Hamnett, D. and Sporton, D. (2012) 'Paying for interviews? Negotiating ethics, power and expectation,' *Area*, 44 (4): 496–502.

Hanson Thiem, C. and Robertson, M. (eds) (2010) Critical Review Forum: 'Behind enemy lines: Reflections on the practice and production of oppositional research,' *Geoforum*, 41 (1): 5–25.

Harding, S. (1987) 'Can there be a feminist method?,' in S. Harding (ed.), *Feminism and Methodology*, Bloomington: Indiana University Press.

McDowell, L. (1992) 'Doing gender: Feminism, feminists and research methods in human geography,' *Transactions of the Institute of British Geographers*, 17 (4): 399–416.

Madge, C. (1993) 'Boundary disputes: Comments on Sidaway (1992),' *Area*, 25 (3): 294–299.

Mistry, J. and Berardi, A. (2012) 'The challenges and opportunities of participatory video in geographical research: Exploring collaboration with indigenous communities in the North Rupununi, Guyana,' *Area*, 44 (1): 110–116.

Rambaldi, G., Kyem, P. A. K., McCall, M. and Weiner, D. (2006) 'Participatory spatial information management and communication in developing countries,' *The Electronic Journal of Information Systems in Developing Countries*, 25: 1–9.

Robson, E. and Willis, K. (eds) (1994) *Postgraduate fieldwork in developing areas: A rough guide*, DARG Monograph No. 8; 2nd edn (1997), DARG Monograph No. 9.

Sidaway, J. (1992) 'In other worlds: On the politics of research by "First World" geographers in the "Third World",' *Area*, 24 (4): 403–408.

Stacey, J. (1988) 'Can there be a feminist ethnography?,' *Women's Studies International Forum*, 11 (1): 21–27.

Steinberg, P. E., Bryan, J., Herlihy, P. H., Cruz, M. and Agnew, J. (2010) Series of short commentaries on the Bowman Expedition, *Political Geography*, 29: 413–423.

Sultana, F. (2007) 'Reflexivity, positionality and participatory ethics: Negotiating fieldwork dilemmas in international research,' *ACME*, 6 (3): 374–385.

Valentine, G. (2005) 'Geography and ethics: moral geographies? Ethical commitment in research and teaching,' *Progress in Human Geography*, 29 (4): 483–487.

Willis, K. (2000) 'No es fácil, pero es posible: The maintenance of middle-class women-headed households in Mexico,' *European Review of Latin American and Caribbean Studies*, 69: 29–45.

Willis, K. (2013) 'Learning from emotions in development fieldwork,' keynote lecture at RidNet [Researchers in Development] Conference: Conducting Fieldwork in Development Contexts, University of Leeds, November 2013.

Willis, K. and Yeoh, B. (1999) 'Gender and transnational household strategies: Singaporean migration to China,' *Regional Studies* 34 (3): 253–264.

Glossary

Terms in *italic* can be found elsewhere in the glossary.

Access The process whereby a researcher makes contact with potential *research participants* (sometimes via a *gatekeeper*), develops *rapport* with them, and negotiates their participation in the research project. Some participants are termed 'difficult to access,' often because their social status, lifestyle, or activities cause them to be hidden from everyday life: for example, illegal immigrants or sex workers.

Critical distance Achieving a balance between immersion in the *research setting* and maintaining a certain level of detachment in order to strive toward *objectivity*. Particularly significant for researchers doing *ethnography* and for those who are *insiders* to the *research community*.

Cross-cultural fieldwork Fieldwork carried out by a researcher in a different culture to their own, usually in a foreign country. In contrast, researchers carrying out fieldwork in a country where they have strong connections (e.g., were born there, hold nationality, or have family ties) are said to be doing fieldwork 'at home,' even if they are no longer resident there.

Cultural sensitivity Encompasses gaining knowledge about the historical, political, social and religious context of the *research setting*; being aware of the rules of social interaction and appropriate behavior; and making attempts to bridge the cultural divide through language learning and other means. Particularly relevant for *outsiders* doing *cross-cultural research*. Cultural sensitivity demonstrates a sense of respect for the *research community* and may assist with developing *rapport*.

Dissemination Communication of research findings: not only to academic audiences through publications and conference presentations, but also to *research participants* and the *research community*, other *stakeholders*, and wider audiences. In the case of policy-relevant research, this involves extracting the key implications of the research findings and communicating them in a way that encourages decision-makers to incorporate the recommendations into policy and practice.

Do no harm A key principle of *ethical research* with human subjects. Researchers should consider whether any of their actions have the potential to harm the lives of the *research participants*, their communities, or their environments in the present or the future, and seek to minimize any such risks. Respecting the *right to privacy* through confidentiality and anonymity is one way of doing this.

Elite research   A study involving *research participants* who are important or influential people in their particular social context, such as politicians, government officials, religious leaders, business executives, academics, organizational directors, or community leaders. When dealing with social elites the researcher's *positionality* and the *power relations* may be quite different from those in interactions with non-elites.

Ethical research   Research which follows basic ethical principles designed to ensure the safety of *research participants* and to prevent irresponsible practices. Key principles of ethical research include *informed consent*, the *right to privacy*, *do no harm*, *justice*, and *reciprocity*. All parts of a research project, from design and data collection to analysis and *dissemination*, should adhere to ethical principles.

Ethics review   A procedure which most researchers have to undertake prior to fieldwork whereby all elements of the research project are evaluated in the light of *ethical research* principles. Researchers are required to describe their proposed *research setting* and *research community*, *sampling procedure*, data collection methods, analysis techniques, and *dissemination* plans, and to demonstrate how they will adhere to ethical research practice. An ethics review board has the authority to approve the proposed field research, request changes to the research design, or reject the proposal.

Ethnography   A *qualitative research* method pioneered in anthropology but now used across the social sciences. It is based on direct observation of a *research community* and its everyday activities in order to understand knowledge, behavior, and meanings. There are two main approaches to ethnography: *participant observation* and non-participant observation. A variety of data collection techniques are usually employed, including keeping a *fieldwork diary* and *interviews*. Ethnographers often engage in *reflexive analysis* to analyze the ways in which their presence in the research community may influence the data collection and research outcomes.

Fieldwork diary   A log kept by a researcher recording the time spent in the field. It may comprise written material as well as photographs, maps, drawings, and other information. There are two types of fieldwork diary. In *ethnography*, keeping a diary is a specific data collection technique whereby the researcher takes systematic and detailed notes describing fieldwork situations and interactions with *research participants*. More generally, some researchers choose to keep a personal diary charting their time in the field, including their changing opinions and emotions through time; while not part of the formal data collection process, such material can be used in *reflexive analysis*.

Focus groups   A qualitative research method which involves organized discussion with a selected group of *research participants* to gain information about their views on and experiences of a particular topic. The researcher acts as a facilitator of the discussion, and participants are encouraged to talk among themselves. This allows the researcher to obtain information through more natural conversation than through *interviews* and also to observe patterns of group interaction. Major advantages of focus groups are that they are fairly low-cost compared to *surveys* and are time-effective compared to individual *interviews*.

Gatekeepers   People from within a *research community*, or having a close relationship with the community, who control the researcher's *access*; for example, a village chief or the senior executive of an organization.

Giving back   See Reciprocity/Giving back.

Global North/Global South   One of the classifications used to divide the countries of the world according to relative levels of development. 'Global North' refers to the richer and more industrialized nations that are largely located in the northern hemisphere, whereas 'Global South' refers to the poorer and less industrialized nations largely located in the southern hemisphere. This broadly corresponds with the classification More Economically Developed Countries (MEDCs) and Less Economically Developed Countries (LEDCs). Both pairs of terms have largely replaced the obsolete classification 'First World'/'Third World' and the terms 'developed countries'/'developing countries.' However, all classifications are generalizations with exceptions and limitations.

Informed consent   A key principle of *ethical research* with human subjects. Informed consent means that the people asked to participate in the research are doing so voluntarily and understand what their participation entails, including their *right to privacy*. Some researchers are required by their institution to prepare a consent form for *research participants* to complete.

Insiders/Outsiders   Researchers who have a direct connection with the *research setting* are termed 'insiders,' while strangers to the context are 'outsiders.' Insider and outsider status is one aspect of researcher identity and *positionality*. While an insider's familiarity with the language and culture, local knowledge, and personal contacts can facilitate the research project, this raises questions about their *subjectivity*. Furthermore, the dichotomy between insider and outsider is not really clear-cut, as many researchers straddle both 'worlds.'

Interviews   A research method involving direct interaction between the researcher and one or more research participants to elicit facts, statements, opinions, and other information. While the spoken words are of primary importance, non-verbal information is also of use to the researcher, including body language, intonation, and silences. There are three main approaches to interviewing. Structured interviews are a form of *survey* whereby each participant is given the same questions in the same order, which allows for the comparison of responses. In semi-structured interviews the researcher follows a guide of questions to ask and topics to cover, while allowing some flexibility in obtaining the information from the respondent. In unstructured interviews there are no predetermined questions and the approach is more informal and conversational. The former approach is used mostly in *quantitative research*, whereas the other two approaches are used in *qualitative research*.

Justice   A key principle of *ethical research* with human subjects. Researchers should make every effort to ensure that no single group disproportionately bears the risks of the research, neither should any single group disproportionately gain the benefits of the research. Justice also relates to the *sampling* procedure, whereby the selection of *research participants* should be done in a fair and equal manner, and also to the special protection of *vulnerable people*.

Key informants   People within the *research community* who, usually due to their social position, have specialist knowledge or access to sources of information that is useful for the researcher, particularly in the early stages of the research project. They may also act as *gatekeepers*.

Objectivity/Subjectivity   Objectivity means being unbiased and not allowing personal beliefs, emotions, or prejudices to influence a situation. However, complete objectivity is impossible in *qualitative research*. All researchers have an inherent subjectivity which influences every aspect of the research project, including the choice of topic, field location and data collection methods, the analysis of data, and the *representation* of findings. Social researchers are encouraged to engage in *reflexive analysis* in order to reflect on how their subjectivity affects the research project.

Outsiders/Insiders   See Insiders/Outsiders.

Participant observation   A *qualitative research* method widely used in *ethnography*, whereby the researcher is embedded in the *research community* and takes part in its activities while also watching and recording them. Some researchers opt for non-participant observation whereby they only observe activities without taking part. Keeping a *fieldwork diary* is one of the main ways in which researchers record their observations.

Participatory research   An approach to doing research which seeks to remove the distinction between the 'researcher' and the 'researched' (or 'expert' and 'beneficiary' in the case of more applied development work). The relationship is collaborative: the research is carried out with, for, or by the *research community*, who are involved in many stages of the research project including defining the problem, designing the study, conducting the data collection, interpreting the findings, and deciding how the results should be used. A participatory approach is considered to be a more ethical means of doing research, as it challenges *power relations*, removes the objectification of *research participants*, and empowers communities. Participatory research methods are often visual and engaging, tailored to participants who may have limited literacy; they include participatory mapping and diagramming; participatory monitoring; video, drama and art; participatory geographical information systems; peer research; and interactive workshops.

Partner organization   An organization with which a researcher collaborates in carrying out the research project; may also act as a *gatekeeper*.

Pilot study   A small-scale study conducted in advance of the full study, commonly used in large-scale *quantitative research*. It may be used to test a hypothesis, a research tool, or aspects of the research design. Findings from the pilot study are used to make adjustments to the research design prior to the full study.

Positionality   A researcher's place in the *research setting*, constructed on the basis of how various aspects of their identity (e.g., age, gender, nationality, educational status) stand in relation to the *research community* and *research participants*. Positionality affects all parts of the research project, including design, *access*, data collection, interpretation, and *representation*.

Power relations   Factors such as educational qualifications, professional status, and wealth can place the researcher in a position of relative power over the research participants, particularly in the context of development research in the *Global South* (although *elite research* can have the opposite effect). Moreover, the researcher is also in charge of the research project, setting its agenda, asking questions, gathering data, and preparing *representation*, which puts them in a position of leadership and dominance. *Participatory research* seeks to challenge power relations and to ensure that the researcher and research participants are in an equal position.

Qualitative research   Method of academic enquiry used across many social science disciplines that aims to gather in-depth understanding of human behavior and communities. Data collection is based on observation and description; methods include *participant observation, focus groups*, semi-structured and unstructured *interviews*, and case studies. Questions tend to be broad and seek to explore 'why' and 'how.' Samples are usually quite small. Data are mostly in word form and are analyzed using techniques such as observer impression, coding, and content analysis.

Quantitative research   Method of academic enquiry used across many social science disciplines that aims to quantify a social phenomenon and looks for trends that can be projected onto a larger population. Data collection is based on measurement and definition; methods include *surveys*, structured *interviews*, and audits. Questions tend to be specific and seek to measure 'what,' 'where,' and 'when.' Samples are usually large. Data are mostly numerical and are analyzed using statistical, mathematical, or computational techniques.

Rapport   A positive relationship that develops between the researcher and *research participants* or the *research community*, involving a sense of mutual trust and emotional affinity; this can enable *access* to participants and the collection of data.

Reciprocity/Giving back   A key principle of *ethical research* with human subjects. The researcher should seek to ensure that the *research participants* and *research community* benefit from taking part in the research. In recognition of their time, information, and resources, the researcher should consider what it is appropriate to give in return: for example, monetary payment, material gifts, or practical help. *Dissemination* is a form of giving back, as it seeks to use the research findings proactively to bring about change.

Reflexivity/Reflexive analysis   A process of self-analysis whereby the researcher reflects on their relationship with the *research setting* and the ways in which their identity and *positionality* may intentionally or unintentionally influence the research process and outcomes. Reflexive writing is an attempt to be transparent about *subjectivity*.

Representation The language used (whether written or oral) by researchers to describe and analyze *research participants* or the data collected from them. The 'politics of representation' refers to the *power relations* embedded within representation (for example, who speaks for whom; whose language is used) and how the representations are subsequently used by the researcher and other *stakeholders*.

Research assistant Someone employed on a temporary basis to help with the research project in some capacity, such as acting as interpreter and/or assisting with data collection. If they are from the *research community*, they may also act as a *gatekeeper*.

Research community The population being studied, from which *research participants* are selected according to a *sampling* procedure.

Research participant Any human subject involved in the research process who provides information to the researcher, whether through *participant observation*, *interviews*, *surveys*, *focus groups*, or other methods. Several terms tend to be used interchangeably, including 'respondent,' 'interviewee,' 'informant,' and 'participant.'

Research setting The context in which research is carried out, which includes the geographical location and socio-political context as well as the *research community*. Research settings vary in scale depending on the subject of study: for example, a village or a country, a workplace or a multinational institution.

Right to privacy A key principle of *ethical research* with human subjects. Researchers must uphold the rights of *research participants* to privacy; this involves making them aware of how the data will be processed, stored, and used (part of the process of *informed consent*) and handling the data with care. Two practices are central to this: confidentiality means limiting access to the data, including storing it securely and not disclosing it to third parties; anonymity provides a way of concealing the identity of research participants in the analysis and presentation of the data.

Risk assessment A procedure which most researchers have to undertake prior to fieldwork, whereby all elements of the research project are evaluated in the light of potential risks to the researcher, the *research participants*, and other *stakeholders*. This includes the researcher's personal safety and awareness of the various hazards in the *research setting*; adherence to the ethical principle that the research should *do no harm* to the research participants; and the responsibilities of an institution in their duty of care to researchers.

Sampling The process of selecting a subset of individuals (e.g., people, organizations) from a *research community*. Information collected from the sample via methods such as *surveys* or *interviews* can be generalized to the entire population of interest. There are a variety of sampling methods to choose from, including random sampling, stratified sampling, systematic sampling, cluster sampling, accidental sampling, and snowball sampling. The researcher's choice of sampling method will depend on a variety of factors including the nature of the population, the availability of information about the population, the need for representativeness, and time-cost concerns.

Stakeholder   Any person, community, or organization which has an interest in the conduct, findings, and outcomes of a research project; these may include *research participants*, the *research community, partner organizations*, host country governments, and international institutions.

Subjectivity/Objectivity   See Objectivity/Subjectivity.

Surveys   A *quantitative research* method used to gather information from *research participants* selected through a *sampling* procedure. From the responses of a representative sample to a predetermined set of questions, the researcher can extrapolate the findings to describe the population as a whole, as well as compare different populations and chart changes over time. Surveys can be divided into two broad types: questionnaires are usually distributed by mail or via the internet and completed by the respondent, while structured *interviews* are carried out by telephone or in person. Both techniques may use closed questions, open-ended questions, or a combination.

Vulnerable people   *Research participants* who are considered to be particularly vulnerable to coercion or undue influence and for whom additional ethical procedures may be implemented as part of the ethical research principle to *do no harm*. These include children, people with disabilities, people with HIV/AIDS, those who are illiterate, the economically disadvantaged, and other socially marginalized groups.

Index

For Product Safety Concerns and Information please contact our EU
representative GPSR@taylorandfrancis.com
Taylor & Francis Verlag GmbH, Kaufingerstraße 24, 80331 München, Germany

www.ingramcontent.com/pod-product-compliance
Ingram Content Group UK Ltd.
Pitfield, Milton Keynes, MK11 3LW, UK
UKHW021013180425
457613UK00020B/924